High Performance Computing and Artificial Intelligence for Geosciences

High Performance Computing and Artificial Intelligence for Geosciences

Editors

Yuzhu Wang
Jinrong Jiang
Yangang Wang

MDPI • Basel • Beijing • Wuhan • Barcelona • Belgrade • Manchester • Tokyo • Cluj • Tianjin

Editors
Yuzhu Wang
China University of Geosciences
Beijing, China

Jinrong Jiang
Chinese Academy of Sciences
Beijing, China

Yangang Wang
Chinese Academy of Sciences
Beijing, China

Editorial Office
MDPI
St. Alban-Anlage 66
4052 Basel, Switzerland

This is a reprint of articles from the Special Issue published online in the open access journal *Applied Sciences* (ISSN 2076-3417) (available at: https://www.mdpi.com/journal/applsci/special_issues/Computing_Artificial_Intelligence_Geosciences).

For citation purposes, cite each article independently as indicated on the article page online and as indicated below:

LastName, A.A.; LastName, B.B.; LastName, C.C. Article Title. *Journal Name* **Year**, *Volume Number*, Page Range.

ISBN 978-3-0365-8180-4 (Hbk)
ISBN 978-3-0365-8181-1 (PDF)

Contents

About the Editors

Yuzhu Wang

Yuzhu Wang received a Ph.D. in engineering from the University of Chinese Academy of Sciences in 2015. He is an associate professor at the School of Information Engineering, China University of Geosciences, Beijing, China. His research interests include high-performance computing, parallel algorithms, and distributed machine learning. He has been involved in several national research projects and has authored over 30 papers in journals and conference proceedings.

Jinrong Jiang

Jinrong Jiang received a Ph.D. in engineering from the Chinese Academy of Sciences in 2007. He is a full professor at the Computer Network Information Center, Chinese Academy of Sciences, Beijing, China. His research interests include parallel algorithms and high-performance computing.

Yangang Wang

Yangang Wang received a Ph.D. in science from Jilin University in 2006. He is a full professor at Computer Network Information Center, Chinese Academy of Sciences, Beijing, China. His research interests include artificial intelligence, parallel algorithms, and high-performance computing.

applied sciences

MDPI

Editorial

High-Performance Computing and Artificial Intelligence for Geosciences

Yuzhu Wang [1,*], Jinrong Jiang [2] and Yangang Wang [2]

1 School of Information Engineering, China University of Geosciences, Beijing 100083, China
2 Computer Network Information Center, Chinese Academy of Sciences, Beijing 100190, China; jjr@sccas.cn (J.J.); wangyg@sccas.cn (Y.W.)
* Correspondence: wangyz@cugb.edu.cn

Citation: Wang, Y.; Jiang, J.; Wang, Y. High-Performance Computing and Artificial Intelligence for Geosciences. *Appl. Sci.* **2023**, *13*, 7952. https://doi.org/10.3390/app13137952

Received: 28 June 2023
Revised: 30 June 2023
Accepted: 2 July 2023
Published: 7 July 2023

1. Introduction

Geoscience, as an interdisciplinary field, is dedicated to revealing the operational mechanisms and evolutionary patterns of the Earth system. Geoscience is very important to enhancing our understanding of the Earth, including its life, resources, and environment. In recent years, the rapid development of high-performance computing and artificial intelligence technologies has presented unparalleled opportunities and challenges to geoscience research. This book aims to explore the significance of high-performance computing and artificial intelligence research in advancing our understanding of geosciences.

High-performance computing provides powerful computational capabilities for geoscience research. The Earth's system encompasses vast amounts of data and intricate interrelationships, including atmospheric dynamics, seismic activities, and ocean circulation, among others. Traditional data processing and modeling methods often struggle to meet the demands of these complex systems. However, the development of high-performance computing technologies has enabled scientists to leverage tools such as parallel computing and supercomputers to process large-scale datasets and conduct more precise simulations and predictions. Through high-performance computing, scientists can enhance their understanding of the dynamical processes within the Earth system, resulting in the improved accuracy and timeliness of weather forecasting, climate simulations, earthquake early warning systems, and more.

Artificial intelligence technology has introduced novel ideas and methodologies into geoscience research. The investigation of geosciences often necessitates the extraction of valuable information from extensive datasets to show the inherent patterns within the Earth system. Machine learning and deep learning algorithms in artificial intelligence are able to process intricate data and discern patterns. They autonomously learn and extract features from massive datasets, aiding scientists in uncovering hidden patterns and trends embedded within the data. For instance, in climate change research, artificial intelligence can analyze climate models and observational data to offer more precise predictions and evaluations. Moreover, artificial intelligence has practical applications in geological exploration and resource surveys, thereby enhancing the efficiency and accuracy of resource exploration.

2. Contributions

The editors acknowledge all contributions, and we are delighted to introduce a collection of eleven selected high-quality research papers in this book.

Qi and colleagues [1] conducted high-performance computing in flood behavior modeling, using the two-dimensional Saint-Venant equations as examples. The equations were discretized using finite difference methods with the explicit leapfrog scheme, considering initial and boundary conditions. They employed MPI, OpenMP, Pthread, and OpenCL to achieve large-scale heterogeneous parallelism and optimized algorithm performance through computation/communication overlap, workgroup optimization, and local memory optimization. Ultimately, their work yielded a well-performing, large-scale, non-uniform parallel solution for the two-dimensional Saint-Venant equations.

Wang and colleagues [2] studied the ZTEM (Z-axis Tipper Electromagnetic Method) electromagnetic field detection method. They performed parallel computations on the inversion algorithm for the two-dimensional ZTEM using MPI. Compared to the serial algorithm, the parallel algorithm achieved acceleration ratios ranging from 1.74 to 3.19 when the number of processes ranged from three to six.

Hao and colleagues [3] conducted a series of parallel optimizations on the LASG/IAP Climate System Ocean Model (LICOM 2.1 version), a high-resolution ocean model independently developed by the Institute of Atmospheric Physics, Chinese Academy of Sciences. The optimizations addressed load imbalance, communication optimization, and loop optimization. Additionally, they employed hybrid parallelization using MPI and OpenMP, as well as asynchronous parallel I/O. The optimized version of LICOM 2.1 achieved more than two-fold acceleration compared to the original version. In large-scale parallel simulations, the optimized version of LICOM scaled up to 245,760 processor cores and resolved the wall time issue during the time integration process.

Yang and coworkers [4] developed site-level gross primary productivity (GPP) using the GeoMAN model, which incorporates spatio-temporal features and incorporates external environmental factors, to predict GPP on the Tibetan Plateau. They evaluated the behavior of four models—random forest (RF), support vector machine (SVM), Deep Belief Network (DBN), and GeoMAN—to predict GPP for nine flux observation sites on the Tibetan Plateau. The GeoMAN model outperformed the other models. These findings are valuable for our understanding of the capability of deep learning models in predicting GPP while aligning with the fundamental knowledge of related fields.

Wang and coworkers [5] proposed a time series prediction model for landslide displacements using mean-based low-rank autoregressive tensor completion (MLATC). They first analyzed the reasons for missing landslide displacement data and designed the corresponding missing dataset. Then, based on the characteristics and internal correlation of landslide displacement monitoring data, they introduced the establishment process of the mean-based low-rank tensor completion prediction model. Finally, they used the proposed method to complete and predict the random missing and non-random missing landslide displacement data. The results of the model were consistent with the original monitoring data and showed good performance in completing and predicting landslide displacement, providing valuable insights for processing missing data and predicting landslide displacement.

Cao and coworkers [6] proposed a new method called Dual Encoder Transform (DualET) for the short-term prediction of photovoltaic (PV) power. The DualET model contained wavelet transform and sequence decomposition blocks for the extraction of information features from image and sequence data, respectively, to improve the correlation of spatial and temporal features. In addition, they proposed a cross-domain attention module to learn the correlation between temporal features and cloud information, and then modified the attention module using alternate forms and Fourier transforms to improve its performance. The model was evaluated on real-world datasets consisting of PV plant data and satellite images, and it outperformed other models in the prediction of short-term PV power generation.

Xu and coworkers [7] used the sliding window method and gray relational analysis to extract features from multi-source real-time monitoring data of landslides in Lishan County, Hunan Province, China. They applied the K-means algorithm with particle swarm optimization for clustering and the Apriori algorithm to mine strong correlation rules between the high-speed deformation process of the landslide and rainfall features. This approach enabled them to identify short-term deformation patterns and precursors of disasters. They indicated that the probability of high-speed deformation of this landslide exceeded 80% when the rainfall occurred within 24 h and the accumulated rainfall in

7 days was greater than 130.60 mm. By using data mining technology to extract short-term deformation patterns of landslides, the accuracy and reliability of early warning systems can be improved.

Du and coworkers [8] analyzed the potential of unsupervised machine learning methods for submarine landslide prediction and compared the performance of three different unsupervised machine learning models (K-means, spectral clustering, and hierarchical clustering) in modeling landslide susceptibility. They selected nine sets of geological factors as input parameters, which were extracted through field investigations. To estimate the susceptibility of submarine landslides, all input factors were grouped into three to four clusters based on data characteristics and environmental variables. The performance of the models was evaluated using internal indicators (the Calinski–Harabasz index, silhouette index, and Davies–Bouldin index) and external indicators (existing landslide distribution, hydrodynamic distribution, and liquefaction distribution) to verify model fit and accuracy. The results showed that all three models (K-means, spectral clustering, and hierarchical clustering) performed well in accurately predicting submarine landslides. Spectral clustering was found to be particularly effective in capturing the environmental geological parameters.

Yang and coworkers [9] proposed an automatic landslide identification method. Their approach combined deep learning with landslide extraction from remote sensing images, using a semantic segmentation model to automate the landslide recognition process. They evaluated the model's performance using metrics from the semantic segmentation task and tested three popular semantic segmentation models (U-Net, DeepLabv3+, and PSPNet) with different backbone networks. The best recognition accuracy of PSPNet was 91.18% mIoU with the classification network ResNet50 as the backbone network, which proved the deep learning method is feasible and effective for use in landslide recognition.

Huang and coworkers [10] proposed a named entity recognition (GNNER) method for geological news based on a bi-directional encoder representation of a converter (BERT) pre-trained language model. The approach addressed drawbacks of traditional word vectors, including the fact that they do not effectively represent the contextual semantics and single extraction effects occur. This approach can also aid the construction of knowledge graphs of geological news. The method involves embedding words in geological news text using a BERT pre-training model, with the resulting word vectors being dynamically obtained and used as input for the model. Next, the word vectors are fed into a bidirectional long- and short-term memory model for further training to obtain contextual features. Finally, the model uses conditional random field sequence decoding to extract six entity types. Through the experiments on the constructed Chinese geological news dataset, the model achieved an average F1 score of 0.839 and was able to recognize news entities in geological news better.

Zhang and coworkers [11] proposed an efficient deep-learning-based mineral identification method, which effectively addressed the limitations of traditional identification methods that heavily rely on the identification capabilities of the identifier and external instruments. The accuracy of existing identification methods is often affected by various factors, including Mohs hardness, color, picture scale, and especially light intensity. Deep-learning-based mineral recognition provides a new solution to this problem, not only saving labor costs but also reducing recognition errors. The authors, using a luminance equalization algorithm, reduced the impact of light intensity on recognition accuracy. First, they proposed a new algorithm combining histogram equalization (HE) and Laplace's algorithm, used the algorithm to process the luminance of the recognized samples, and finally used the YOLOv5 model to recognize the samples and implemented a deep learning mineral recognition method based on luminance equalization.

3. Concluding Remarks

We believe that this book will provide inspiration for researchers who are studying high-performance computing or AI methods for geosciences. In the future, the research proposed in this book could be further optimized to improve its effectiveness and allow it to be applied in other similar applications.

Conflicts of Interest: The authors declare no conflict of interest.

References

1. Qi, Y.; Li, Q.; Zhao, Z.; Zhang, J.; Gao, L.; Yuan, W.; Lu, Z.; Nie, N.; Shang, X.; Tao, S. Heterogeneous Parallel Implementation of Large-Scale Numerical Simulation of Saint-Venant Equations. *Appl. Sci.* **2022**, *12*, 5671. [CrossRef]
2. Wang, M.; Tan, H.; Wang, Y.; Lin, C.; Peng, M. Parallel Computation for Inversion Algorithm of 2D ZTEM. *Appl. Sci.* **2022**, *12*, 12664. [CrossRef]
3. Hao, H.; Jiang, J.; Wang, T.; Liu, H.; Lin, P.; Zhang, Z.; Niu, B. Deep Parallel Optimizations on an LASG/IAP Climate System Ocean Model and Its Large-Scale Parallelization. *Appl. Sci.* **2023**, *13*, 2690. [CrossRef]
4. Yang, Q.; Nie, N.; Wang, Y.; Wu, X.; Liu, W.; Ren, X.; Wang, Z.; Wan, M.; Cao, R. Spatial–Temporal Correlation Considering Environmental Factor Fusion for Estimating Gross Primary Productivity in Tibetan Grasslands. *Appl. Sci.* **2023**, *13*, 6290. [CrossRef]
5. Wang, C.; Zhao, Y. Time Series Prediction Model of Landslide Displacement Using Mean-Based Low-Rank Autoregressive Tensor Completion. *Appl. Sci.* **2023**, *13*, 5214. [CrossRef]
6. Cao, H.; Yang, J.; Zhao, X.; Yao, T.; Wang, J.; He, H.; Wang, Y. Dual-Encoder Transformer for Short-Term Photovoltaic Power Prediction Using Satellite Remote-Sensing Data. *Appl. Sci.* **2023**, *13*, 1908. [CrossRef]
7. Xu, J.; Bai, D.; He, H.; Luo, J.; Lu, G. Disaster Precursor Identification and Early Warning of the Lishanyuan Landslide Based on Association Rule Mining. *Appl. Sci.* **2022**, *12*, 12836. [CrossRef]
8. Du, X.; Sun, Y.; Song, Y.; Xiu, Z.; Su, Z. Submarine Landslide Susceptibility and Spatial Distribution Using Different Unsupervised Machine Learning Models. *Appl. Sci.* **2022**, *12*, 10544. [CrossRef]
9. Yang, S.; Wang, Y.; Wang, P.; Mu, J.; Jiao, S.; Zhao, X.; Wang, Z.; Wang, K.; Zhu, Y. Automatic Identification of Landslides Based on Deep Learning. *Appl. Sci.* **2022**, *12*, 8153. [CrossRef]
10. Huang, C.; Wang, Y.; Yu, Y.; Hao, Y.; Liu, Y.; Zhao, X. Chinese Named Entity Recognition of Geological News Based on BERT Model. *Appl. Sci.* **2022**, *12*, 7708. [CrossRef]
11. Zhang, J.; Gao, Q.; Luo, H.; Long, T. Mineral Identification Based on Deep Learning Using Image Luminance Equalization. *Appl. Sci.* **2022**, *12*, 7055. [CrossRef]

applied sciences

MDPI

Article

Heterogeneous Parallel Implementation of Large-Scale Numerical Simulation of Saint-Venant Equations

Yongmeng Qi [1], Qiang Li [1,*], Zhigang Zhao [1], Jiahua Zhang [1], Lingyun Gao [2], Wu Yuan [2], Zhonghua Lu [2], Ningming Nie [2], Xiaomin Shang [1] and Shunan Tao [1]

1 School of Computer Science and Technology, Qingdao University, Qingdao 266071, China; 2019025931@qdu.edu.cn (Y.Q.); zgzhao@qdu.edu.cn (Z.Z.); zhangjh@radi.ac.cn (J.Z.); 2020025815@qdu.edu.cn (X.S.); 2021023842@qdu.edu.cn (S.T.)

2 Computer Network Information Center Chinese Academy of Sciences, Beijing 100083, China; gaolingyun@cnic.cn (L.G.); yuanwu@sccas.cn (W.Y.); zhlu@sccas.cn (Z.L.); nienm@sccas.cn (N.N.)

* Correspondence: chucklee@qdu.edu.cn; Tel.: +86-18669818931

Abstract: Large-scale floods are one of the major events that impact the national economy and people's livelihood every year during the flood season. Predicting the factors of flood evolution is a worldwide problem. We use the two-dimensional Saint-Venant equations as an example and for high-performance computing in modelling the flood behavior. Discretization of the two-dimensional Saint-Venant equations with initial and boundary conditions with the finite difference method in the explicit leapfrog scheme is carried out. Afterwards, we employed a large-scale heterogeneous parallel solution on the "SunRising-1" supercomputer system using MPI, OpenMP, Pthread, and OpenCL runtime libraries. On this basis, we applied communication/calculation overlapping and the local memory acceleration to optimize the performance. Finally, various performance tests of the parallel scheme are carried out from different perspectives. We have found this method is efficient and recommend this approach be used in solving systems of partial differential equations similar to the Saint-Venant equations.

Keywords: Saint-Venant equations; finite difference method; parallel computing; heterogeneous computing

Citation: Qi, Y.; Li, Q.; Zhao, Z.; Zhang, J.; Gao, L.; Yuan, W.; Lu, Z.; Nie, N.; Shang, X.; Tao, S. Heterogeneous Parallel Implementation of Large-Scale Numerical Simulation of Saint-Venant Equations. *Appl. Sci.* **2022**, *12*, 5671. https://doi.org/10.3390/app12115671

Academic Editors: Daniel Dias and Yuzhu Wang

Received: 25 April 2022
Accepted: 31 May 2022
Published: 2 June 2022

1. Introduction

Reservoir dams have played a huge role in flood control, water supply, power generation, etc. [1], and are one of the important components of the hydraulic engineering system. But if a dam is damaged due to various factors, the consequences are generally catastrophic. Excessive flooding is one of the natural causes of dam failure. Floods are the most frequent natural disasters, affecting large numbers of people and agricultural lands, as well as causing casualties and damage to infrastructure. Increased runoff rates due to urbanization, prolonged rainfall, and insufficient river capacity are some of the main causes of flooding. After entering the flood season in 2020, there have been multiple rounds of heavy rainfall in southern China, resulting in severe floods in many places. As of 9 July 2020, flood disasters had affected 30.2 million people in 27 provinces (district and cities), with direct economic losses of 61.79 billion yuan. In 2021, Henan, Shanxi, and other places were also hit by heavy rainstorms. Therefore, understanding the behavior of water flow in a channel is critical for early flood disaster management and saving lives. The research direction of this paper is to study the behavior of flood through numerical simulation of surface water flow, and to perform heterogeneous parallel processing and visualization of the simulation process.

Mathematical models are scientific or engineering models constructed using mathematical logic methods and mathematical language to solve various practical problems. In hydraulic science, mathematical models are often used to simulate fluids of different phenomena, and numerical methods are used to control the fluid models. What we use in our research is the flood wave propagation dynamics equation, which was first proposed by Saint-Venant [2], so it is also called the Saint-Venant (SV) equations. It consists of a continuity equation reflecting the law of conservation of mass and an equation of motion reflecting the law of conservation of momentum, and is widely used to predict surface flow parameters such as velocity, depth, or height. Nowadays, they are used to model flows in a wide variety of physical phenomena, such as overland flow, flooding, dam breaks, tsunami [3]. SV equation was originally only used to describe one-dimensional surface water flow, for two-dimensional, SV equation is derived from the Navier-Stokes equation [4], and two-dimensional SV equation is often referred to as the shallow water equation(SWE). Since SV equations are mathematically quasi-linear hyperbolic partial differential equations (PDE), it is difficult to obtain the analytical solution of SV equations by analytical method. Therefore, different numerical methods have been proposed to simulate surface water flow.

Long before the advent of computers, some people solved the SV equations through numerical simulation. Reinaldo and Rene [5] long ago used the explicit MacCormack time-splitting scheme to establish a mathematical model for solving two-dimensional SV equations. Two industrial applications at the time are also presented, demonstrating the validity of the model. Later, Fiedler and Ramirez [6] also used this method to simulate a discontinuous two-dimensional hydrodynamic surface flow equation (a variant of the two-dimensional SV equations) with spatially variable properties. The method was developed to model spatially variable infiltration and microtopography, and can also be used to model irrigation, tidal flat and wetland cycles, and flooding. With the development of computers and the improvement of computing performance, more and more people put the simulation process on the computer. For example, Kamboh [7] et al. also established a mathematical model with initial conditions and boundary conditions using two-dimensional Saint-Venant PDE in order to predict and simulate flood behavior. Next, the corresponding models are discretized and implemented on MATLAB using the common explicit finite difference method. The finally generated graph structure can visually see the changes of each parameter over time. Asier [8] also used the two-dimensional Saint-Venant equation to simulate precipitation or runoff events, but the method used was the finite volume method. He developed and compared the following three programming methods: sequential, multi-threaded and many-core architectures. The multi-threaded code is written using OpenMP and the many-core architecture is written using CUDA. He concluded that the performance of the GPU parallel version using CUDA is strongly affected by the size of the problem. It is also proposed that combining MPI and GPU methods can improve computational efficiency and data capacity, but this is not implemented in the paper. In the field of heterogeneous computing, Ding [9] et al. transplanted, parallelized, and accelerated the solver of the one-dimensional S-V equation based on MPI and athread library. The athread library is an accelerated thread library designed for the master-slave acceleration programming model of the SW26010 processor. They use MPI to realize the parallelization between the master cores, and use athread to accelerate the slave cores. After that, optimization methods such as SIMD (Single Instruction Multiple Data) vectorization and communication/calculation overlapping were carried out. In addition to the above work, adaptive mesh refinement (AMR) is also an important part of the algorithm performance. AMR is generally efficient and effective in treating problems with multiple spatial and temporal scales. AMR improves the quality of solution on a mesh by refining cells only in places where a high grid resolution is desired, thereby increasing the memory efficiency and computation speed [10]. Xin Zhao [11] proposed a 3D volume-of-fluid method based on the adaptive mesh refinement technique. He introduced projection methods on adaptive grids to solve the incompressible Navier–Stokes equations. In order to simulate ocean wave propagation, Michael [12] et al. proposed a method for numerical simulation of dynamically adaptive problems on adaptive triangular grids with recursive structure. They used a grid generation process based on recursive bisection of triangles along marked edges to achieve 2D dynamically adaptive discretization. Kevin and Frank [13] used a multilayer lattice Boltzmann model (LBM) to solve the 3D wind-driven shallow water flow problems, and studied the performance of the parallel LBM for the multilayer shallow water equations on the CPU-based high performance computing environment using OpenMP. It is concluded that explicit loop control with cache optimization in LBM provides better performance than the implicit loop control on execution time, speedup, and efficiency as the number of processors increases. There are many more applications of the two-dimensional Saint-Venant equation, such as [14–17].

There are already some frameworks for solving PED using GPUs or accelerated devices. For example, Bhadke [18] et al. used CUDA to develop a 3D-CFD computing framework for the conduction process. They discretized the model into a three-dimensional grid and solved it using an alternating-direction implicit method. In summary, although many different studies have modeled the SV equation (e.g., [19,20]), and there are discussions on parallelization and performance optimization, there are few studies on large-scale Saint-Venant systems. Therefore, this research aims to build a two-dimensional simple finite difference model and use MPI, OpenMP, Pthread, and OpenCL for heterogeneous large-scale processing. In this work, we first introduce the governing equation and calculation method of SV equations. Then we introduce the basic implementation of our heterogeneous massively parallel computing. Finally, the parallel strategy is optimized and the performance is tested.

2. Governing Equations and Numerical Method

The governing equations used in our research are as follows:

$$\frac{\partial z}{\partial t}+\frac{\partial(zu)}{\partial x}+\frac{\partial(zv)}{\partial y}=0 \tag{1}$$

$$\frac{\partial(zu)}{\partial t}+\frac{\partial\left(zu^2+\frac{gz^2}{2}\right)}{\partial x}+\frac{\partial(zuv)}{\partial y}=gz\left(S_{0x}-S_{fx}\right) \tag{2}$$

$$\frac{\partial(zv)}{\partial t}+\frac{\partial(zuv)}{\partial x}+\frac{\partial\left(zv^2+\frac{gz^2}{2}\right)}{\partial y}=gz\left(S_{0x}-S_{fx}\right) \tag{3}$$

Equation (1) is derived from the conservation of mass, and Equations (2) and (3) are derived from the conservation of momentum in the x and y directions, respectively. Among them, z refers to the elevation (depth or height) of the water flow in the open channel, u and v are the water velocity in the x and y directions respectively, t is the time, g is the acceleration of gravity. S_0 and S_f refer to the water surface gradient and frictional resistance. In order to use this system of equations more conveniently in the computer, we need to use the product rule of differentiation to further simplify the system of equations. We then discretize the equations using an explicit finite difference scheme, where the time and space derivatives are respectively discretized by the following expressions:

$$\frac{\partial u}{\partial t}\approx\frac{u_{i,j}^{k+1}-u_{i,j}^{k-1}}{2\Delta t},\ \frac{\partial u}{\partial x}\approx\frac{u_{i+1,j}^{k}-u_{i-1,j}^{k}}{2\Delta x},\ \frac{\partial u}{\partial y}\approx\frac{u_{i,j+1}^{k}-u_{i,j-1}^{k}}{2\Delta y}$$

This format is the central difference format, in which the central difference in time is also called the leapfrog format, and its advantage is that it can enhance the stability of the calculation. In this way, the discretized equations become the following form:

$$\frac{z_{i,j}^{k+1}-z_{i,j}^{k-1}}{2\Delta t}+u_{i,j}^{k}\frac{z_{i+1,j}^{k}-z_{i-1,j}^{k}}{2\Delta x}+z_{i,j}^{k}\frac{u_{i+1,j}^{k}-u_{i-1,j}^{k}}{2\Delta x}+v_{i,j}^{k}\frac{z_{i,j+1}^{k}-z_{i,j-1}^{k}}{2\Delta y}+z_{i,j}^{k}\frac{v_{i,j+1}^{k}-v_{i,j-1}^{k}}{2\Delta y}=0 \tag{4}$$

$$\begin{aligned}&u_{i,j}^{k}\frac{z_{i,j}^{k+1}-z_{i,j}^{k-1}}{2\Delta t}+z_{i,j}^{k}\frac{u_{i,j}^{k+1}-u_{i,j}^{k-1}}{2\Delta t}+\left(u_{i,j}^{k}\right)^2\frac{z_{i+1,j}^{k}-z_{i-1,j}^{k}}{2\Delta x}+2z_{i,j}^{k}u_{i,j}^{k}\frac{u_{i+1,j}^{k}-u_{i-1,j}^{k}}{2\Delta x}+gz_{i,j}^{k}\frac{z_{i+1,j}^{k}-z_{i-1,j}^{k}}{2\Delta x}+\\&u_{i,j}^{k}v_{i,j}^{k}\frac{z_{i,j+1}^{k}-z_{i,j-1}^{k}}{2\Delta y}+z_{i,j}^{k}v_{i,j}^{k}\frac{u_{i,j+1}^{k}-u_{i,j-1}^{k}}{2\Delta y}+z_{i,j}^{k}u_{i,j}^{k}\frac{v_{i,j+1}^{k}-v_{i,j-1}^{k}}{2\Delta y}=gz_{i,j}^{k}\left(S_{0x}-S_{fx}\right)\end{aligned} \tag{5}$$

$$v_{i,j}^{k}\frac{z_{i,j}^{k+1}-z_{i,j}^{k-1}}{2\Delta t}+z_{i,j}^{k}\frac{v_{i,j}^{k+1}-v_{i,j}^{k-1}}{2\Delta t}+\left(v_{i,j}^{k}\right)^{2}\frac{z_{i,j+1}^{k}-z_{i,j-1}^{k}}{2\Delta y}+2z_{i,j}^{k}v_{i,j}^{k}\frac{v_{i,j+1}^{k}-v_{i,j-1}^{k}}{2\Delta y}+gz_{i,j}^{k}\frac{z_{i,j+1}^{k}-z_{i,j-1}^{k}}{2\Delta y}+ \\ u_{i,j}^{k}v_{i,j}^{k}\frac{z_{i+1,j}^{k}-z_{i-1,j}^{k}}{2\Delta x}+z_{i,j}^{k}v_{i,j}^{k}\frac{u_{i+1,j}^{k}-u_{i-1,j}^{k}}{2\Delta x}+z_{i,j}^{k}u_{i,j}^{k}\frac{v_{i+1,j}^{k}-v_{i-1,j}^{k}}{2\Delta x}=gz_{i,j}^{k}\left(S_{0y}-S_{fy}\right) \tag{6}$$

Then, after simply shifting the term and removing the denominator, the three variables z, u, and v can be solved iteratively. As shown in Figure 1, each time an iteration value is calculated, the results of the previous two iterations are used, which is also a feature of the leapfrog format. Because of this, the stability of the calculation process is strengthened.

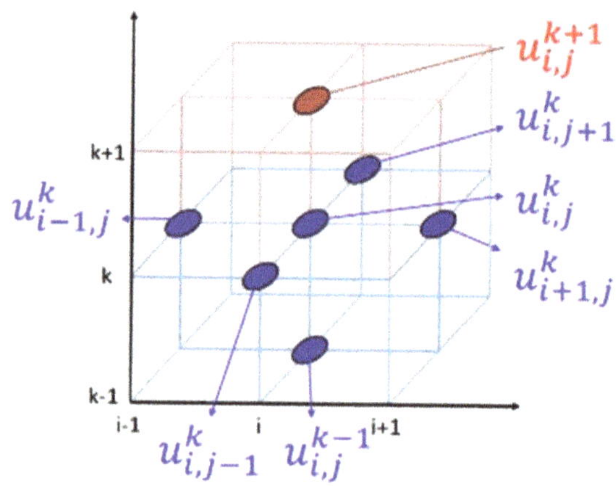

Figure 1. The value used by the iterative process.

It is also because of the solution value range shown in Figure 1 that the boundary of the grid will be exceeded when calculating the edge value. At this time, we need to introduce boundary conditions and add a circle of ghost cells around the standard grid. Balzano [21] gives a general review of existing wetting and drying (WD) methods. He gave expressions for boundary conditions at moving boundaries, which can be solved on fixed grids, adaptive grids and moving grids based on natural coordinates. Moreover, a number of solutions or models for dealing with boundary problems based on this method are introduced. Finally, a discussion on implicit finite-difference models is carried out. Heniche [22] et al. also gave specific boundary conditions based on the WD model. They proposed two kinds of boundary conditions: solid boundary and open boundary. Specific conditions must be imposed when dealing with solid boundaries, while open boundaries are used to specify flow regimes. The basic idea of the boundary conditions we use is to bounce incoming particles toward the boundary back into the fluid [23]. In order to achieve this boundary condition, it is only necessary to make the boundary value equal to the edge value.

After that, showing the initial conditions, we consider using a rectangular area with no bottom stress and wind stress. The water surface at each location is stationary and has a height of 10 and the flow is zero, i.e., z = 10 m, u = 0 m/s, v = 0 m/s. Then a flood wave with a maximum height of 1 m is generated at the entrance of the water to simulate the flow behavior of fluid in a single channel, through the formula:

$$z=e^{-\frac{(a-a_0)^2+(b-b_0)^2}{k^2}}$$

In this formula, a_0 and b_0 are the location of the highest flood wave in the x and y directions respectively, where a_0 = 25, b_0 = 1 is taken to locate the center of the entrance. k is the initial height of the water wave, where k = 10 m. As for a and b, they are the coordinates of each grid node on the coordinate axis. This will get the highest value when $a = a_0$, $b = b_0$. The initial state of a small-scale simulation process is shown in Figure 2.

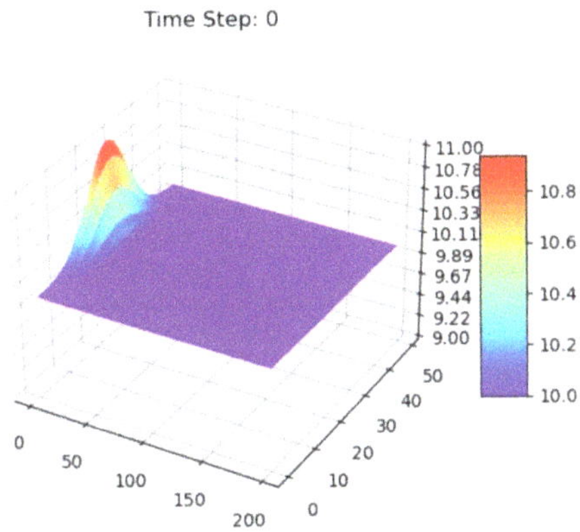

Figure 2. Schematic of a small-scale initial state.

3. Heterogeneous Implementation

First of all, our heterogeneous implementation works together through the CPU and accelerator. The CPU also participates in the calculation process while being responsible for communication, while the accelerator is only responsible for calculation. After that, we used MPI, OpenMP, Pthread, and OpenCL runtime libraries. Among them, MPI is a parallel program interface based on multiple processes with good performance, which is used in this paper for point-to-point communication between nodes. Both OpenMP and Pthread are thread-parallel interfaces. OpenMP is a portable API that is very convenient to use because it does not bind the program to a pre-set thread. But because of this, we cannot use OpenMP to manage specific threads individually. So, we introduced Pthread. The Pthread API handles most of the behavior required by threads. These behaviors include creating and terminating threads, waiting for threads to complete, and managing interactions between threads. Combining the advantages and disadvantages between the two, we use OpenMP when performing simple thread parallelism, and use Pthread when more complex thread operations are required. OpenCL is the first open, free standard for general-purpose parallel programming for heterogeneous systems. It is widely used in parallel processors such as multi-core CPUs, GPUs [24–26], accelerators, etc. We are here to process the accelerator and take care of some of the computing tasks, through OpenCL.

A heterogeneous parallel framework can be used across multiple platforms, but different clusters are equipped with different equipment, and the program will be changed accordingly. This experiment is performed on the domestic advanced computing system "SunRising-1". The specific experimental environment is shown in Table 1.

Table 1. SunRising-1 experimental environment.

Hard/Software Environment	Name	Details/Version
Hard environment (single node)	CPU	32-core domestic × 86 processor × 1
	RAM	16 GB DDR4 × 8
	Acceleration device	Domestic GPGPU accelerator × 4, 16 GB HBM2 VRAM, bandwidth 1 TB/s
	Network	InfiniBand HDR network, Fat-tree topology, 200 Gbps

Table 1. *Cont.*

Hard/Software Environment	Name	Details/Version
Software environment	MPI	Openmpi 4.0.4
	gcc/g++	4.8.5
	OpenMP	3.1
	Pthread	NPTL 2.17
	OpenCL	Platform: AMD Accelerated Parallel Processing Driver version: 2982.0 OpenCL Standard: OpenCL 2.0

Figure 3 shows the schematic of our heterogeneous parallel framework. We first initialize MPI and use MPI_Type_contiguous() to create an MPI datatype by replicating an existing datatype (for example, MPI_INT, MPI_DOUBLE, etc.). These replications are created into contiguous locations, resulting in a contiguous datatype created. The created datatype is then committed using MPI_Type_commit(), before it can be used for communication. After entering the process parallelism, routinely initialize OpenCL, including obtaining the platform, device, creating context, etc. The next step is to implement the initial value conditions described in Section 2, and we call this process "initialization". The cluster used in our experiment is equipped with 32 CPU cores and four accelerators for a single node, so for the initialization process, we enable four threads to simultaneously call OpenCL to start the kernel function. This process does not involve complicated operations, so OpenMP is used. Thread parallelism ends after the kernel function finishes running. After that the calculation process starts. In this process, both the CPU and the accelerator participate in the calculation, so we divided two thread groups, one of which contains four threads to perform operations similar to the previous process to start the kernel function, and the other is to enable the remaining 28 threads participate in the computation. The grouping and waiting of threads are involved here, so Pthread are used. After an iterative calculation is over, because of the existence of ghost cells, communication is required next. Because of the large scale of computation, a node may need to exchange data bidirectionally with four adjacent nodes (maybe 2 or 3 times for nodes at the edge). We use four MPI_Sendrecv() functions to implement this communication operation. MPI_Sendrecv() combines sending a message to a destination and receiving a message from another process into one call. Figure 4 shows how MPI_Sendrecv() in the four directions exchanges data. For the intermediate nodes, each node must send data to the surrounding adjacent nodes, and also receive data from the surrounding adjacent nodes, which can be easily implemented by calling MPI_Sendrecv().

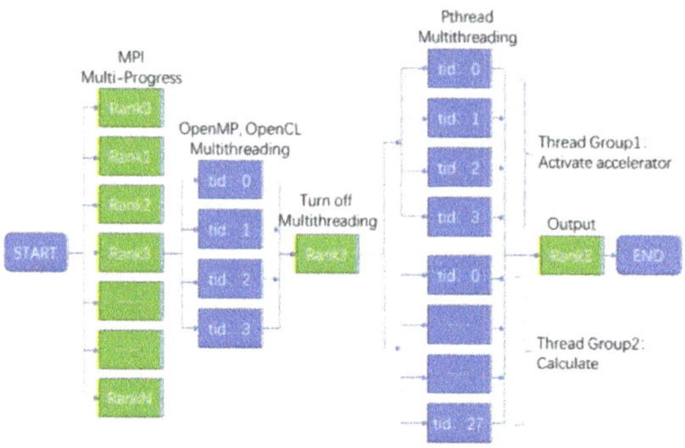

Figure 3. Schematic of the heterogeneous parallel framework.

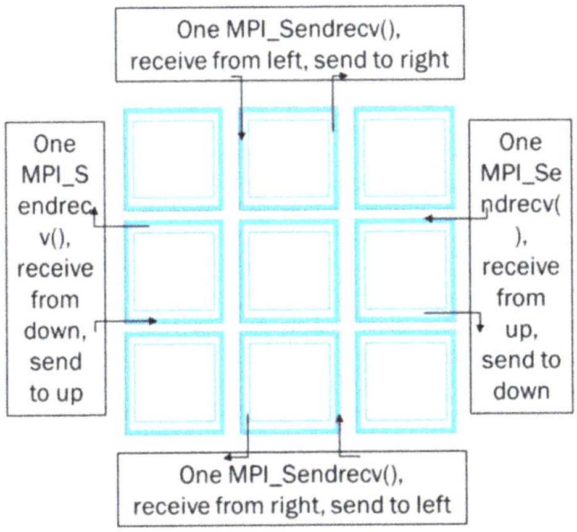

Figure 4. Schematic of data exchange.

When dealing with nodes at edges, we introduce MPI_PROC_NULL, a constant that represents a dummy MPI process rank. This allows all MPI processes to issue the same calls regardless of their position.

After the communication ends, a complete iteration is over, and after that, the iterative loop continues until the set time is reached. Then we output the result and perform post-processing to generate a picture, such as the visualization of the output of the z value shown in Figure 5 (Only the parts with numerical changes are displayed). In this way, you can intuitively see the change of the water level in the river channel with time.

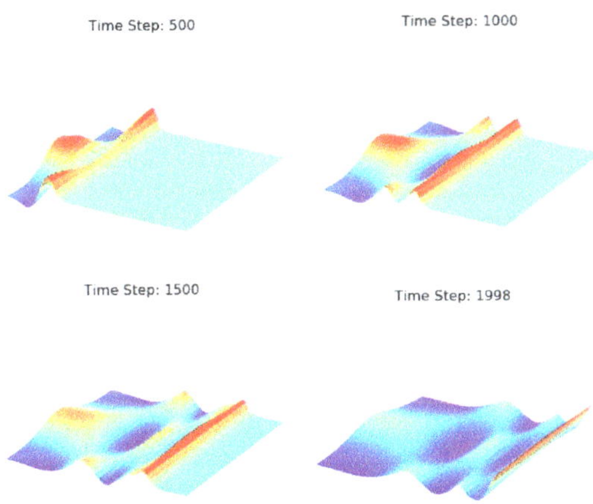

Figure 5. Simulation of water surface elevation at different time steps.

4. Performance Optimization and Testing

In this section, we will optimize the current parallel strategy and test its performance.

4.1. Overlap Computation and Communication

Scaling MPI applications on large high performance computing systems requires efficient communication patterns. Whenever possible, the application needs to overlap communication and computation to hide the communication latency. For the current parallel model, we consider implementing an efficient ghost cell exchange mechanism using non-blocking, peer-to-peer communication (mainly the MPI_Isend() and MPI_Irecv() functions), and domain decomposition of the grid.

First, we must confirm how to decompose the domain, that is, to confirm which part of the value to be calculated does not involve MPI communication. The grid we use is composed of real values inside and a dummy value of an outer circle of ghost cells, where real values are computed and dummy values are communicated. It can be seen from Figure 1 that the value of the ghost cell is only used when calculating the value of the outermost circle of the actual value. Therefore, as shown in Figure 6, we can come up with a solution for the domain decomposition. We call the outermost circle of the actual value the "halo cell", and the part of the actual value that removes the halo is called the "inner filed".

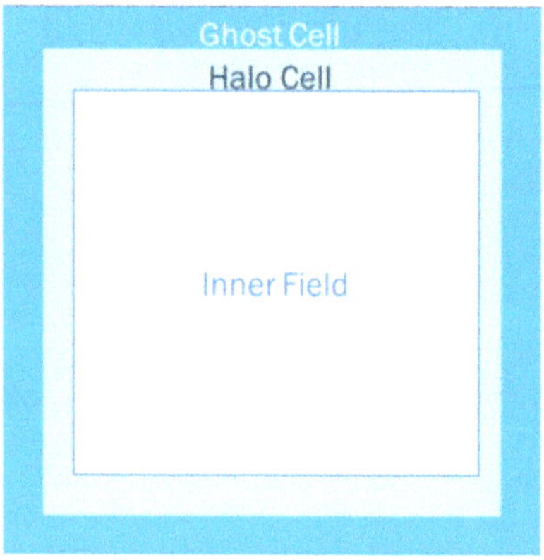

Figure 6. Schematic of domain decomposition.

The general iterative scheme is summarized as follows:

1. Copy data of the ghost cells to send buffers;
2. Ghost cell exchange with Isend/Irecv calls;
3. Compute (part 1): Update the inner field of the domain;
4. Call MPI_Waitall();
5. Copy data from receive buffers to the ghost cells;
6. Compute (part 2): Update the halo cells;
7. Repeat.

After completing the calculation and communication overlap, we performed a comparison test with the previous version, as shown in Table 2. The calculation amount of a single node is 200 iterations, and the grid size in each node is 2564×4100.

Table 2. Comparison test before and after overlap.

Version	Calculating Time (s)	Sync Time (s)	Total Time (s)
Before	240.61	6	246.61
After	189	5	194

In this test, we performed a total of 200 iterations, and the time obtained after the average of five times was tested. The total time simply refers to the time to execute 200 iterations, excluding the previous initialization and subsequent output time. The sync time is the time to wait with MPI_Barrier() before the end of each iteration. It can be seen that by overlapping communication and calculation, the communication process is hidden in the calculation process, which can greatly reduce the time consumption, and even when the calculation amount is large enough, the communication time can be completely ignored.

Afterwards, we conducted extended tests on this overlapped version. We tested the parallel performance of the program by continuously increasing the number of processes while keeping the amount of computation allocated by each process basically the same. The test results are shown in the Figure 7. This test uses four nodes (processes) as the benchmark, and these four nodes are arranged in a 2×2 manner. The number of iterations and grid size are the same as before. It can be seen that the overlapping of communication computing can not only shorten the computing time but also maintain a good parallel efficiency.

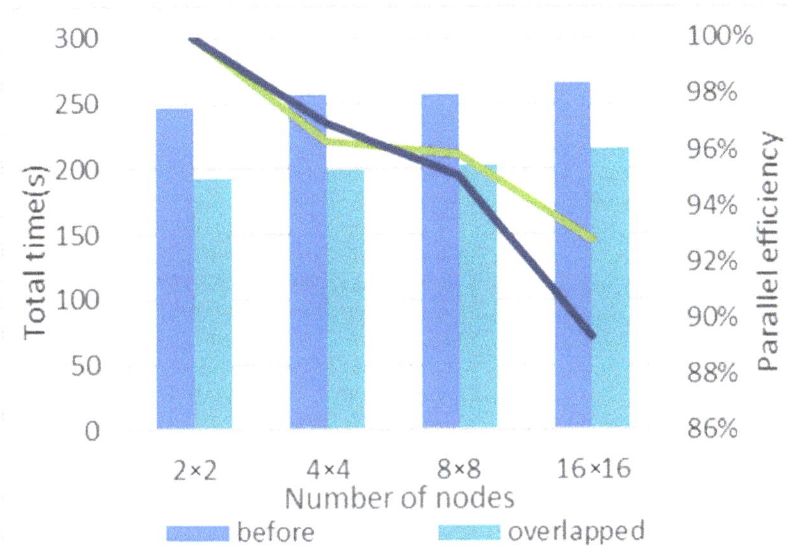

Figure 7. Extended test performance graph for overlapped version.

4.2. Work Group Optimization

First of all, a few definitions are briefly described. Each OpenCL device has one or more compute units, and each compute unit is composed of one or more processing elements. The basic unit of executing a kernel in OpenCL is called a work-item, and a collection of several work-items is called a work-group. A work-group executes on a single compute unit. The work-items in a given work-group execute concurrently on the processing elements of a single compute unit. There are two ways to specify the number of work-groups. In the explicit model a programmer defines the total number of work-items to execute in parallel and also how the work-items are divided among work-groups. In the implicit model, a programmer specifies only the total number of work-items to execute in parallel, and the division into work-groups is managed by the OpenCL implementation. We used the implicit model before, and the optimization method is to use the explicit model. In order to choose an appropriate work-group size, we conducted tests and the results are shown in the Figure 8. Finally, the default maximum work-group size is different for different devices. For example, the device limit we use is 256, which is the same as most devices. At this point, in the kernel code, add __attribute__((amdgpu_flat_work_group_size(<min>,<max>))) after the kernel function so that the kernel can be launched on when the working group is greater than 256. It can be seen from the test results that the best results can be achieved when the work-group size is 256.

Figure 8. Tests for different workgroup sizes.

4.3. Using Local Memory

Using Local Memory is related to OpenCL's memory model. The Local Memory is the memory belonging to a certain computing unit. The host cannot see or operate on this part of the memory. This area allows all processing elements inside the computing unit to read and write, and these processing elements can share it. This also means that this is a memory area associated with a workgroup and can only be accessed by work items in that workgroup. Local Memory is the smallest unit that can be shared in the OpenCL memory structure, so making full use of Local Memory is a deep and very effective optimization method.

After fully understanding the OpenCL memory structure and the importance of using Local Memory, we start from the storage and calculation methods of the current parallel scheme, and continue to use the topology in the previous section to design the Local Memory usage scheme of this program. When this program uses the OpenCL device for calculation, it transfers the custom structure array of three iterations to the Global Memory. The first two iterations store the data that has been calculated before, and the third iteration is to store the data that will be calculated. Here, resmat (result matrix) is used to represent the array of the three iterations, resmat[0–2] represents the three iterations respectively, resmat[0] represents the first iteration, and so on to obtain the representation of the other two iterations. By analyzing the calculation formula after the simple deformation of Equations (4)–(6), the frequency of use of resmat[1] has reached 64 times, resmat[0] has 5 times, and resmat[2] also has 5 times. This means that to complete an iterative calculation, the Global Memory needs to be accessed at least 74 times in a single device, and frequent reading and writing of the Global Memory is bound to affect the performance of the program. Therefore, the optimization idea for this program is to store the data in resmat[1], which has the highest frequency of accessing the Global Memory, into the Local Memory of each work group according to the topology.

There is also a prerequisite for Local Memory optimization, that is, the size of Local Memory is limited, and the increase in read rate is achieved by sacrificing capacity. In order to ensure that the envisaged solution can be executed, we obtain the device information through the clGetDeviceInfo function, and by setting the cl_device_info parameter to CL_DEVICE_LOCAL_MEM_SIZE, the size of the Local Memory area can be obtained. In the experimental environment used by this program, the size of the Local Memory is 65,536 bytes (B), which is 64 kilobytes (KB). It can be seen that this capacity is very limited, so it is necessary to control the size of the data passed into the Local Memory.

After there are hardware limitations, go back to the Local Memory usage scheme of this program. In order to use a piece of continuous data to facilitate writing to Local Memory and the limitation of Local Memory space, only resmat[1] is considered here, and resmat[0] and resmat[2] are no longer considered. Since the topology in the previous section is used, and grid nodes and work items are in one-to-one correspondence, a work group only undertakes the computing task of a row of 256 grid nodes. When calculating this part of the task, in addition to the data of the corresponding position in resmat[1], the data of the four adjacent positions of each grid node will be used, which can be seen in Figure 1. Therefore, processing similar to ghost cell is required, that is, the grid data of 3 × 258 scale is transferred to Local Memory. Each grid node consists of three double-type variables z, u, and v. Each double-type occupies 8 bytes in the system environment. Therefore, the space occupied by grid data of 3 × 258 scale is 18,576 B, which satisfies the limitations of OpenCL devices.

After the scheme is designed, it is the specific implementation. First, a new parameter with the __local qualifier must be added to the definition of the kernel function. This parameter is usually a pointer to a memory space, and then clSetKernelArg is used in the main function to set the parameter. At the same time, the specified size must be the size allocated by the __local parameter, and the parameter value must be NULL, because the host cannot access the Local Memory, so the data in the Global Memory can only be copied to the Local Memory in the kernel function. A set of index functions provided by OpenCL is used to determine the position of a work item in its own work group and the position in the global, so that the data in this position in the Global Memory can be copied to the corresponding position in the Local Memory. Finally, after the above processing methods, the original resmat[1] accesses the Global Memory 64 times into a single Global Memory access and 64 local memory accesses. The test results of the final program are shown in Figure 9.

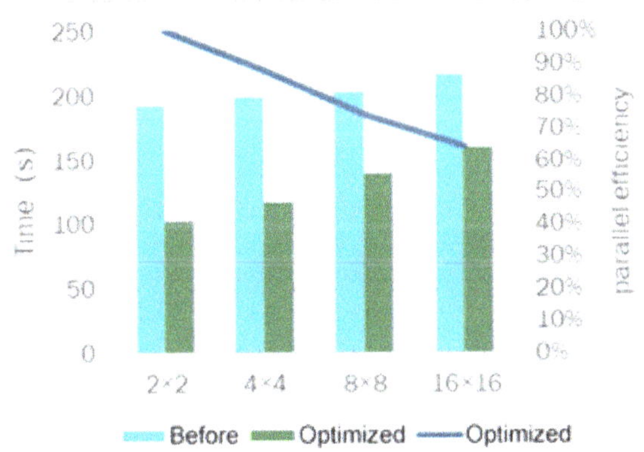

Figure 9. Extended test performance graph for overlapped version.

It can be seen that the running time is greatly reduced after using Local Memory, especially in the case of 2 × 2 nodes, the running time is optimized to 1/2 of the original. At the same time, the rate of decline in parallel efficiency has slightly increased. This is because the calculation time has been optimized, and there will be no major changes as the number of computing nodes increases. However, the original synchronization time, thread opening, and other times remain basically unchanged compared with those before optimization, and will also change with the increase of computing nodes. As the number of computing nodes increases, the proportion of computing time in the total time becomes lower, which affects the parallel efficiency, so the increase in the speed of parallel efficiency decline is within reasonable expectations.

After that, we further tested the optimization scheme. In the current scheme, the local memory usage size is 18,576 B, and a large part of the space is still unused compared to the limit of 65,536 B. In this solution, the size of the working group determines the amount of Local Memory usage. Therefore, we combined two optimization methods for testing. We started with 128 work items and increased in units of 128 to test the performance of the program under different work group sizes and using different sizes of Local Memory. The results are shown in Table 3.

Table 3. Comparison of test results for different workgroup sizes.

Workgroup Size	Use Local Memory Size (B)	Test Results (s)
128	9360	84.36
256	18,576	83.85
384	27,792	84.27
512	37,008	84.64
640	46,224	84.16
768	55,440	84.30
896	64,656	84.30
1024	73,872	Error

As can be seen from the table, in the case of 1024 work items, an error is reported because the size of the Local Memory used exceeds the limit of 65,536 B. In addition, the rest of the results are not very different. In OpenCL, logically all threads are parallel, but in reality, not all threads can execute at the same time from a hardware perspective, but multiple thread groups are scheduled through the hardware's own scheduling algorithm. The thread group is the smallest execution unit that is scheduled in the acceleration device defined by each hardware manufacturer. For example, on the NVIDIA CUDA platform, this thread group is called warp and consists of 32 threads, and on the AMD platform and in this lab environment, they are called wavefronts and consist of 64 threads. How many such thread groups can be executed at the same time is determined by the number or size of the Local Memory, cache, registers, and SIMD (Single Instruction, Multiple Data) instruction set of the computing unit. Therefore, under large-scale computing tasks, the total number of tasks executed at the same time is roughly the same. Therefore, it can be concluded that the size of the work group has little effect on the final result when the computing scale is large and the computing tasks are close to saturation.

Finally, we also conduct further tests on the NVIDIA platform to verify the conclusions, and the test results are shown in Table 4. This test was carried out on a single-node NVIDIA GeForce RTX 3090 platform. Due to the memory limitation of the platform, the grid node size of this test was 32×1536, and 100,000 iterations were calculated and the time was counted. It can be seen that the size of the workgroup does not affect the final result on the NVIDIA platform. Here, the NVIDIA 3090 graphics card has been tested to find that its Local Memory limit is 49,152 B, and when the workgroup size is 640 and the Local Memory size is 46,224 B, the reason for still reporting an error is that the CL_INVALID_WORK_GROUP_SIZE error in OpenCL is triggered, that is the maximum workgroup size of the OpenCL device is exceeded.

Table 4. Comparison test results of different workgroup sizes under the NVIDIA platform.

Workgroup Size	Use Local Memory Size (B)	Test Results (s)
128	9360	37.11
256	18,576	37.37
384	27,792	37.75
512	37,008	36.33
640	46,224	Error

4.4. Heterogeneous Parallel Performance Testing

First, let us take a look at the performance of heterogeneous parallelism without any optimization. In the charts in this section, the bar charts represent time and the line charts represent parallel efficiency. Only weak scaling tests are performed here, as shown in Figure 10. Because the performance of the weak extension test is not good and the parallel efficiency declines too fast, this paper does not perform multiple tests and strong extension tests on this version. After analysis, it is found that a small calculation scale is used here, as mentioned above, such problems are greatly affected by the problem size. Then we scaled up and performed the optimizations mentioned in the previous section. Figures 11 and 12 are the results of the weak and strong scaling test after optimization.

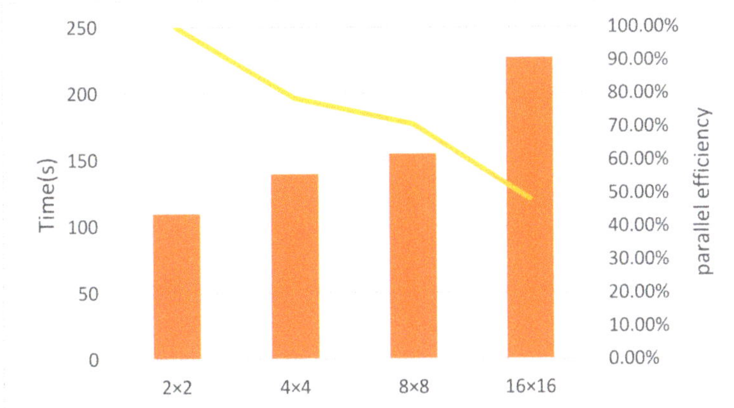

Figure 10. Unoptimized heterogeneous parallel weak scaling test.

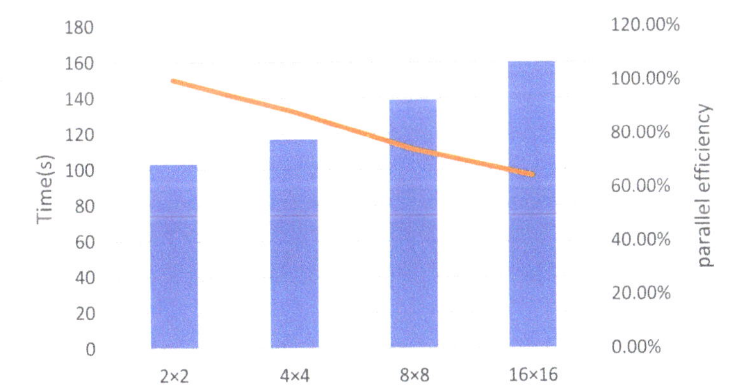

Figure 11. Optimized heterogeneous parallel weak scaling test.

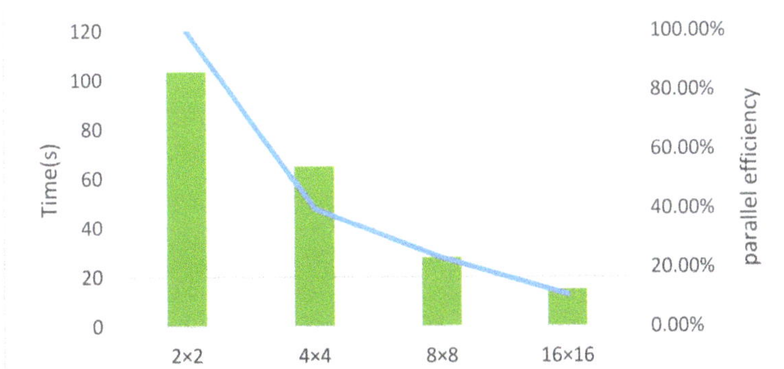

Figure 12. Optimized heterogeneous parallel strong scaling test.

These tests were performed on the same scale as the previous overlap optimization. It can be seen that this version of the weak scaling test can still have good parallel efficiency when the calculation time is greatly reduced. The performance of the strong scaling test is not good, because the calculation time is optimized and decreases with the number of nodes, while the synchronization time increases with the increase of computing nodes. This leads to a lower proportion of computing time in the total time, which affects the parallel efficiency. So, the increase in the rate of parallel efficiency decline is reasonably expected.

After that, the total scale is expanded to obtain better parallel efficiency. This time, the scale under 2×2 nodes is $10{,}244 \times 16{,}388$, and the scale under 16×16 nodes is 1284×2052. The test results are shown in Figure 13, it can be seen that the parallel efficiency is improved after the total scale is enlarged. In order to further obtain better parallel efficiency, the calculation scale can continue to be scaled up, but it does not make practical sense to do so.

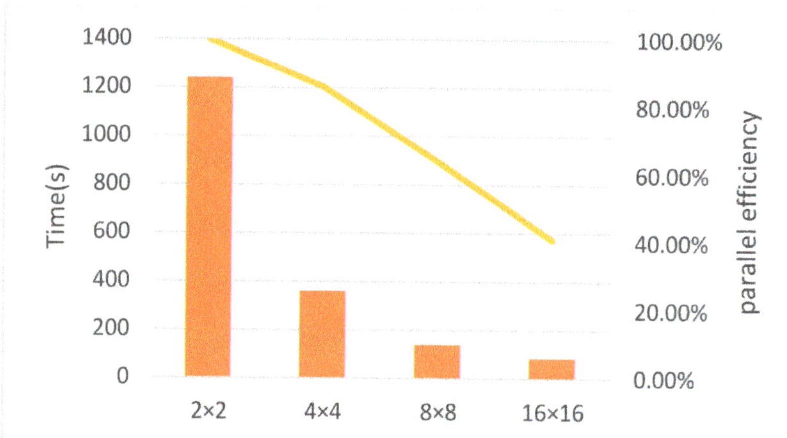

Figure 13. Optimized heterogeneous parallel strong scaling test. In the first section, we mentioned that Bhadke et al. also proposed a computing framework for solving PDE using GPU. However, according to our calculation method of parallel efficiency, the parallel efficiency calculated according to the speedup results provided by them is not ideal. For example, the speedup ratio calculated based on the grid size of 1000 ($10 \times 10 \times 10$) in their article, the speedup ratio also increases with the increase of the grid size, but the parallel efficiency is only below 30%.

4.5. Computational Time Comparison Tests

In this section, we conduct various performance comparison tests for CPU and accelerator respectively.

4.5.1. CPU Single-Core and Multi-Core Comparison

The first is to test the acceleration performance of the CPU using Pthread compared to the single-core serial mode. In our program, Pthread is used to launch 28 threads to be responsible for some calculations, and its acceleration performance is shown in the Figure 14. The horizontal axis is the calculation scale. We first fix the length to expand the width, and then fix the width to expand the length. It can be seen that no matter what the expansion is, the speedup is stable at around 3. Therefore, using Pthread can not only achieve more complex thread operations, but also obtain stable acceleration effects.

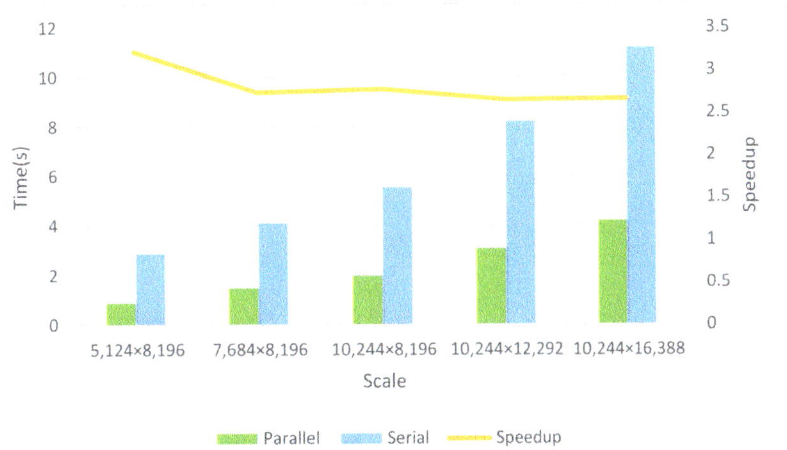

Figure 14. Single-core CPU vs multi-core CPU comparison chart.

4.5.2. CPU and Accelerators Comparison

We then compared CPU and accelerators that perform computations simultaneously. Because the cluster used in our experiment is equipped with 32 CPU cores and 4 accelerators for a single node, we did two sets of tests, one is to compare a single 28-thread CPU with a single accelerator under the same scale, and the other is to compare four accelerators with a single 28-thread CPU under the same scale. Figures 15 and 16 are the result graphs of the above two tests. It can be seen that the acceleration effect of the accelerator is very obvious, which is why a good parallel solution must use GPU or accelerator. Heterogeneous computing can make full use of the performance of CPUs and acceleration devices, reflecting that heterogeneous computing will be the future development direction of parallel computing.

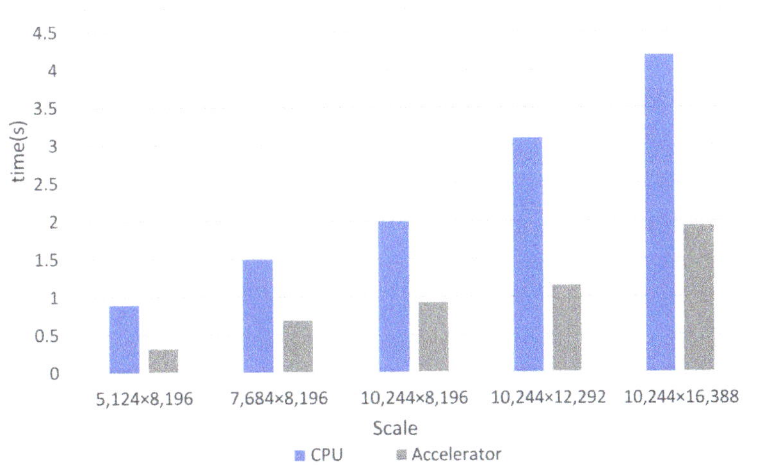

Figure 15. Single CPU vs. single accelerator comparison chart.

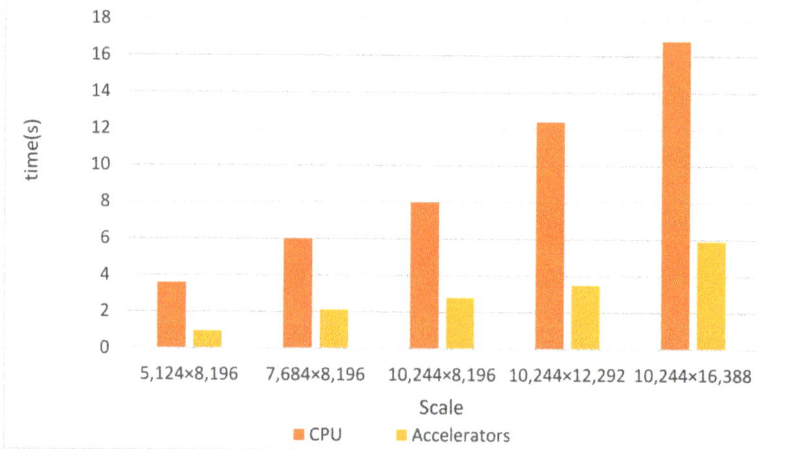

Figure 16. Single CPU vs. 4 accelerators comparison chart.

5. Conclusions

In this work, two-dimensional Saint-Venant equations were implemented by using the leapfrog-style finite difference method for the purpose of studying flood behavior. Large-scale heterogeneous parallel solution is implemented using MPI, OpenMP, Pthread, and OpenCL runtime libraries. The heterogeneous strategy is optimized by overlapping communication calculation, setting up work groups, etc., and performance tests from various perspectives are also carried out. Finally, a large-scale heterogeneous parallel solution for 2D SV equations with good performance is obtained. In the future, we hope that our work can become a large-scale heterogeneous parallel solution framework capable of solving SV equations -like PDE. For example, the heat conduction equation as an important PDE should also be able to use our computational framework. It describes how the temperature in a region changes with time, and like the Saint-Venant equation, it is difficult to obtain analytical solutions, and numerical methods are usually used to obtain numerical solutions. There are also some other problems that may also be able to use our computational framework, such as phase transitions, elasticity, electrical potential, etc. The purpose of our current research is to propose a massively heterogeneous parallel framework based on Saint-Venant's equations in domestic advanced computing systems. Therefore, a simple model is used to facilitate heterogeneous parallel implementation. In the future work, we will consider more complex and more types of models to simulate various actual situations (such as floods, tsunamis, dam failures, etc.).

Author Contributions: Conceptualization, Q.L. and N.N.; methodology, Q.L. and N.N.; software, Y.Q.; validation, L.G., W.Y. and Z.L.; formal analysis, Q.L.; investigation, Y.Q.; resources, Q.L.; data curation, Y.Q. and X.S.; writing—original draft preparation, Y.Q.; writing—review and editing, Q.L., Z.Z. and J.Z.; visualization, Y.Q. and S.T.; supervision, Q.L.; project administration, N.N. All authors have read and agreed to the published version of the manuscript.

Funding: This research was funded by the National Key R&D Program of China (Grant No. 2020YFB1709501), GHFund A (No. 20210701) and the Shandong Province Natural Science Foundation (Grant No. ZR201910310143).

Institutional Review Board Statement: Not applicable.

Informed Consent Statement: Not applicable.

Conflicts of Interest: The authors declare no conflict of interest.

References

1. Xiang, Y.; Sheng, J.B.; Yang, M.; Zhang, S.C.; Yang, Z.H. Impacts on ecological environment due to dam removal or decommissioning. *Chin. J. Geotech. Eng.* **2008**, *30*, 1758.
2. Saint-Venant, B.D. Theory of unsteady water flow, with application to river floods and to propagation of tides in river channels. *Fr. Acad. Sci.* **1871**, *73*, 148–154.
3. Hu, J. A simple numerical scheme for the 2D shallow-water system. *arXiv* **2017**, arXiv:1801.07441.
4. Dawson, C.; Mirabito, C.M. The Shallow Water Equations. 2008. Available online: https://users.oden.utexas.edu/~{}arbogast/cam397/dawson_v2.pdf (accessed on 20 April 2022).
5. Garcia, R.; Kahawita, R.A. Numerical solution of the st. venant equations with the maccormack fi-nite-difference scheme. *Int. J. Numer. Methods Fluids* **2010**, *6*, 259–274. [CrossRef]
6. Fiedler, F.R.; Ramirez, J.A. A numerical method for simulating discontinuous shallow flow over an infiltrating surface. *Int. J. Numer. Methods Fluids* **2000**, *32*, 219–239. [CrossRef]
7. Kamboh, S.A.; Sarbini, I.N.; Labadin, J.; Eze, M.O. Simulation of 2D Saint-Venant equations in open channel by using MATLAB. *J. IT Asia* **2016**, *5*, 15–22. [CrossRef]
8. Lacasta, A.; Hernández, M.M.; Murillo, J.; García-Navarro, P. GPU implementation of the 2D shallow water equations for the simulation of rainfall/runoff events. *Environ. Earth Sci.* **2015**, *74*, 7295–7305. [CrossRef]
9. Ding, Z.-Z.; Chu, G.-S.; Hu, C.-J.; Li, Y. Parallelization and optimization of Saint-Venant solver on Sunway many-core processor. *Comput. Eng. Sci.* **2020**, *43*, 820–829.
10. Huang, X. Adaptive mesh refinement for computational aeroacoustics. In Proceedings of the 11th AIAA/CEAS Aeroacoustics Conference, Monterey, CA, USA, 23–25 May 2005.
11. Zhao, X. A three-dimensional robust volume-of-fluid solver based on the adaptive mesh refinement. *Theor. Appl. Mech. Lett.* **2021**, *11*, 100309. [CrossRef]
12. Bader, M.; Böck, C.; Schwaiger, J.; Vigh, C. Dynamically Adaptive Simulations with Minimal Memory Requirement—Solving the Shallow Water Equations Using Sierpinski Curves. *SIAM J. Sci. Comput.* **2010**, *32*, 212–228. [CrossRef]
13. Tubbs, K.R.; Tsai, T.C. Multilayer shallow water flow using lattice boltzmann method with high performance computing. *Adv. Water Resour.* **2009**, *32*, 1767–1776. [CrossRef]
14. Esteves, M.; Faucher, X.; Galle, S.; Vauclin, M. Overland flow and infiltration modelling for small plots during unsteady rain: Numerical results versus observed values. *J. Hydrol.* **2000**, *228*, 265–282. [CrossRef]
15. Valiani, A.; Caleffi, V.; Zanni, A. Case Study: Malpasset Dam-Break Simulation using a Two-Dimensional Finite Volume Method. *J. Hydraul. Eng.* **2002**, *128*, 460–472. [CrossRef]
16. Caleffi, V.; Valiani, A.; Zanni, A. Finite volume method for simulating extreme flood events in natural channels. *J. Hydraul. Res.* **2003**, *41*, 167–177. [CrossRef]
17. Kim, D.-H.; Cho, Y.-S.; Yi, Y.-K. Propagation and run-up of nearshore tsunamis with HLLC approximate Riemann solver. *Ocean Eng.* **2007**, *34*, 1164–1173. [CrossRef]
18. Bhadke, Y.; Kawale, M. Development of 3D-CFD code for heat conduction process using CUDA. In Proceedings of the 2014 International Conference on Advances in Engineering and Technology Research, ICAETR, Unnao, India, 1–2 August 2014. [CrossRef]
19. Di Cristo, C.; Greco, M.; Iervolino, M.; Martino, R.; Vacca, A. A remark on finite volume methods for 2D shallow water equations over irregular bottom topography. *J. Hydraul. Res.* **2020**, *59*, 337–344. [CrossRef]
20. Altaie, H.; Dreyfuss, P. Numerical solutions for 2D depth-averaged shallow water equations. *Int. Math. Forum* **2018**, *13*, 79–90. [CrossRef]
21. Balzano, A. Evaluation of methods for numerical simulation of wetting and drying in shallow water flow models. *Coast. Eng.* **1998**, *34*, 83–107. [CrossRef]
22. Heniche, M.; Secretan, Y.; Boudreau, P.; Leclerc, M. A two-dimensional finite element drying-wetting shallow water model for rivers and estuaries. *Adv. Water Resour.* **2000**, *23*, 359–372. [CrossRef]
23. Yang, Z.; Bai, F.; Xiang, K. A lattice Boltzmann model for the open channel flows described by the Saint-Venant equations. *R. Soc. Open Sci.* **2019**, *6*, 190439. [CrossRef]
24. Wang, Y.; Guo, M.; Zhao, Y.; Jiang, J. GPUs-RRTMG_LW: High-efficient and scalable computing for a longwave radiative transfer model on multiple GPUs. *J. Supercomput.* **2020**, *77*, 4698–4717. [CrossRef]
25. Wang, Y.; Zhao, Y.; Jiang, J.; Zhang, H. A Novel GPU-Based Acceleration Algorithm for a Longwave Radiative Transfer Model. *Appl. Sci.* **2020**, *10*, 649. [CrossRef]
26. Wang, Y.; Zhao, Y.; Li, W.; Jiang, J.; Ji, X.; Zomaya, A.Y. Using a GPU to Accelerate a Longwave Radiative Transfer Model with Efficient CUDA-Based Methods. *Appl. Sci.* **2019**, *9*, 4039. [CrossRef]

applied sciences

MDPI

Article

Parallel Computation for Inversion Algorithm of 2D ZTEM

Mao Wang [1], Handong Tan [2,*], Yuzhu Wang [1], Changhong Lin [2] and Miao Peng [2]

[1] School of Information Engineering, China University of Geosciences, Beijing 100083, China
[2] School of Geophysics and Information Technology, China University of Geosciences, Beijing 100083, China
* Correspondence: thd@cugb.edu.cn

Abstract: ZTEM refers to the Z-axis tipper electromagnetic method. The ZTEM method is an airborne magnetotelluric sounding method based on the difference in rocks' resistivity using the native electromagnetic field. The method is effective in exploring large-scale structures when the ground is fluctuant. The paper introduces the inversion algorithm of 2D ZTEM named the conjugate gradient method. This method, which avoids solving the Jacobi matrix, is very effective but not effective enough when the model is divided into a big grid. This study can perform further computation using parallel computation and then receive the processed data. We compare the results of the serial algorithm with the result of the parallel algorithm, which proves that the parallel algorithm is correct. When the number of processes is between three and six, the speedup ratio is between 1.74 and 3.19. It improves the effectiveness of the parallel algorithm of 2D ZTEM.

Keywords: ZTEM; 2D forward modeling; inversion; parallel algorithm; tipper

Citation: Wang, M.; Tan, H.; Wang, Y.; Lin, C.; Peng, M. Parallel Computation for Inversion Algorithm of 2D ZTEM. *Appl. Sci.* **2022**, *12*, 12664. https://doi.org/10.3390/app122412664

Academic Editor: Jianbo Gao

Received: 9 November 2022
Accepted: 5 December 2022
Published: 10 December 2022

1. Introduction

The ZTEM (Z-axis tipper electromagnetic method) method is a native electromagnetic field exploration method. The ZTEM method is developed based on the MT. The abbreviation of magnetotelluric is MT. We can collect magnetic data in a helicopter using the ZTEM method. The method has special advantages in mineral deposit exploration and environmental exploration [1–3]. We need to spend a great deal of time collecting MT data on the ground when the MT method is applied to explore the area. If we can collect data in a helicopter, we can save more time and money. It is very difficult to collect data on the electric field, but tipper data are available from a helicopter. The tipper is the ratio of the vertical magnetic field and horizontal magnetic field.

With the development of this instrument, MT data processing technology has matured. The ZTEM method became possible. Domestic and foreign researchers have conducted research on the ZTEM inversion. Xu Zhibo performed research on 2D ZTEM inversion [4], Li ziqiang performed research on 3D ZTEM [5], Gu performed research on a parallel algorithm of 3D MT inversion [6], Holtham conducted research on 3D ZTEM inversion [7], and Sasaki conducted research on 3D ZTEM [8]. The research on the parallel algorithm of 2D ZTEM inversion is limited, so it is essential to perform research on the parallel algorithm of 2D ZTEM inversion. Two-dimensional conjugate gradient inversion is an effective method. The model is divided into m columns and n rows. When the number is large, more computation time is required. The model is divided into 400 × 400 in this paper, and the run time is 9804 s. We need to develop a parallel algorithm to solve the problem. First, we introduce the parallel algorithm.

The rest of this paper is organized as follows. Section 2 introduces 2D modeling and inversion. Section 3 introduces the MPI technology and how to realize the parallel algorithm. Section 4 shows the results and analysis of the parallel algorithm. Section 5 summarizes this paper and proposes an outlook for future work.

2. 2D ZTEM Forward Modeling and Inversion

2.1. ZTEM Forward Modeling

According to the convention of the 2D ZTEM problem, the earth is regarded as a Half space conductor ($z \geq 0$) and the insulated air layer is above the earth. The electromagnetic field source is above the ground ($z = -j$). The ZTEM vertical field is collected in different positions using the MT method, and the horizontal field is the same as in the MT method.

(1) Math-physical model

The Maxwell equation differential form is expressed as:

$$\nabla \times E = i\omega\mu_0 H \tag{1}$$

$$\nabla \times H = \sigma E \tag{2}$$

Usually, a Descartes rectangular coordinate system is used. The origin of the coordinate is on the ground, the z-axis is straight down, and the y-axis is horizontal. E is the electric field, H is the magnetic field, μ_0 is the magnetic conductivity, σ is the conductivity, and ω is the angular frequency. There are two polarization models in Maxwell equations.

We set the height of the helicopter as h, and the positive z-axis is down to the earth in the coordinate system. The tipper T is as follows:

$$H_z = T_{zx}H_x + T_{zy}H_y \tag{3}$$

Because the H_z is only in the TE mode, Equation (3) is simplified as:

$$H_z = T_{zy}H_y \tag{4}$$

(2) The finite differential method is used to solve the 2D ZTEM forward model problem

We use the finite differential method to determine the 2D ZTEM forward modeling problem. The main work of the forward problem is to solve the partial differential equation. We set the boundary condition of the equation, and we can obtain the result of the problem.

$$Kv = s \tag{5}$$

K is the coefficient matrix. v is the magnetic field. s is the right vector of the equation. We solve the equation and obtain the magnetic field. Finally, we can obtain the tipper.

2.2. 2D Inversion Modeling

The forward model is as follows:

$$d = F(m) + e \tag{6}$$

d is the data vector, m is the model vector, e is the error vector, and F is the forward function [9–11].

The data vector $d = \left[d^1, d^2, \cdots, d^N\right]^T$, d^i is the tipper in an observation point. The model vector is $m = \left[m^1, m^2, \cdots, m^M\right]^T$. The number of the grid is M, and m^j is the resistivity of the unit.

The objective function is as follows:

$$\psi(m) = (d - F(m))^T V^{-1}(d - F(m)) + \lambda m^T L^T L m \tag{7}$$

λ is the regularization parameter; V is the variance of error vector e; the matrix L is the Laplace operator. The first item is the data-fitting difference, and the second item is the model-smoothing item [12–14].

Because a great deal of time is spent computing the Jacobi matrix, in order to save run time, the non-linear conjugate gradient inversion algorithm avoids computing the Jacobi matrix. It computes the product of the Jacobi matrix (or the Jacobi transpose matrix) and a vector x. We need to perform two pseudo-forwards and then obtain $A^T V^{-1} e$ and Ap. Figure 1 shows the flow chart of inversion.

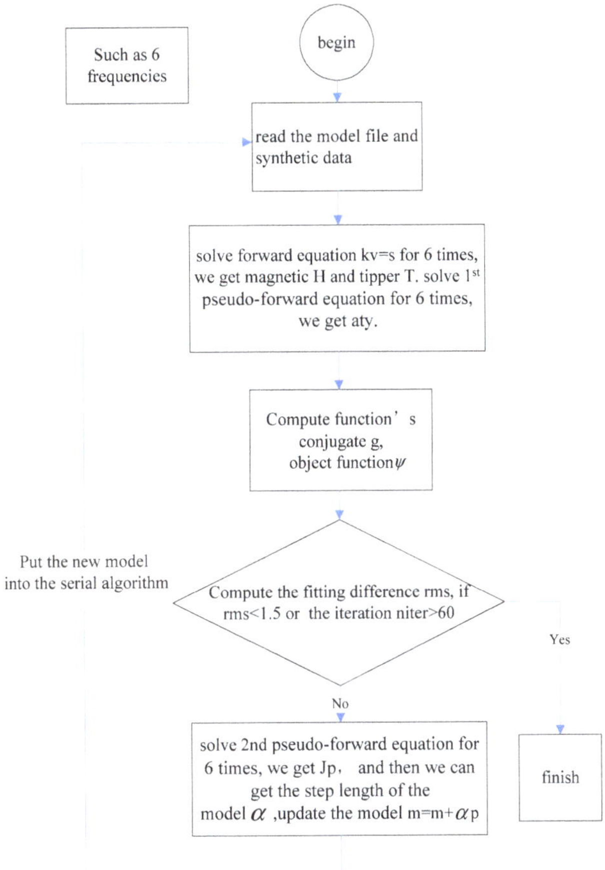

Figure 1. The flow chart of inversion.

This requires a great deal of time to compute the conjugate of the object function $g = -2A^T V^{-1} e + 2\lambda L^T Lm$ and $f = Ap$. When we obtain the value of $A^T V^{-1} e$ and Ap [15], we can obtain the step length α and update the model $m = m + \alpha p$.

3. Parallel Algorithm

3.1. MPI Technology

We can realize data casting, sending, receiving, and synchronizing. The MPI supports several data types [16], including complex. The MPI program is one program, but every process can perform different computations. Every process has an ID, and we can allocate tasks to every process by ID.

The variable's name is the same, but the values of the variables in different processes can be different. If the processes want to share the data, they need to communicate with each other.

3.2. Parallel Algorithm of ZTEM Inversion

The algorithm needs to perform the calculation many times. A calculation has three parts, namely, solving the equation forward (obtaining the value of T), the first pseudo-forward solution to the equation of inversion (obtaining the value of Aty), and the second pseudo-forward solution to the equation of inversion (obtaining the value of Ap). We analyzed the algorithms of forward and inversion and found that solving the equation is the main work and costs a large proportion of the running time. As with six frequencies, when the frequency is different, we can obtain the k, and k is also different. K is in the equation $kv = s$. In order to obtain the new m, we need to solve the equation 18 times in an iteration process. When there are three processes in the algorithm, every process needs to solve the equation six times. The process needs to solve the equation 18 times in the serial algorithm. The process in the parallel program needs to do less work than the serial program, so the parallel program needs less time to execute the inversion algorithm than the serial program. Therefore, the parallel efficiency is higher [17].

The six frequencies are freq(1) ... freq(6) 400, 200, 100, 50, 20, and 10 Hz. When the frequency is higher than other frequencies, solving the equation needs less time. Two kinds of processes exist in the algorithm. One is the main process and the other is the subprocess. The 0 process is the main process. It owns the global data, distributes the work, sends the data, receives the data, and writes the result files. The subprocesses receive data from the 0 process, perform the work, and send the result to the 0 process. When there are three processes in the parallel program, the 0 process is the main process and the first and second processes are the subprocesses. The 0 process also needs to perform the work in the algorithm. We will introduce the main procedure.

(1) MPI_INIT(), In the MPI environment. The main process reads the file, obtains the frequencies, models the information, and observes the data. MPI_Bcast() is the main process, which broadcasts the data to the other processes. The beginning resistivity of the object model is set as the background resistivity.
(2) In Table 1, the allocated frequencies and process ID information are shown. Every process needs to solve the equation of the relevant frequencies separately. We obtain the apparent Tipper of the observed point after solving the forward equation. We obtain the $A^T V^{-1} e$ (the symbol is named aty_mpi in the program). In MPI_recv(), the main process obtains the moderestm_all and aty_all of other processes. Aty_all has six vectors of different frequencyies. Aty is the sum of six vectors. Mode_restm_all has six modrestm_mpi, and the sequence is 1, 2, 3, 4, 5, and 6. Then, we obtain the modrestm to continue computing.

Table 1. The allocated frequencies.

Process ID	The Frequency
0	freq(1), freq(2)
1	freq(3), freq(4)
2	freq(5), freq(6)

In MPI_Bcast(), the main process broadcasts the data, including m and Aty, to the other processes.

(3) After computing the value of object function ψ, the main process can obtain the rms. If rms is less than 1.5, we proceed to step 4; otherwise, we compute the second pseudo-forward equation. Every process needs to perform computing separately. Process can obtain the Ap_mpi. MPI_recv(), and the main process receives Ap_mpi from the other processes. The process puts it together and obtains the value of Ap_all. Ap_all has six Ap_mpi, of which the order is 1, 2, 3, 4, 5, and 6. We obtain the Ap. In MPI_Bcast(), the main process broadcasts Ap to the subprocesses, and we obtain the value of step length α. We can obtain the new model $m = m + \alpha p$, go to step 2.

(4) The main process writes the *m* data in the result file. In MPI_FINALIZE(), we finish the parallel environment. The flow chart of the parallel computing of 2D inversion is shown in Figure 2.

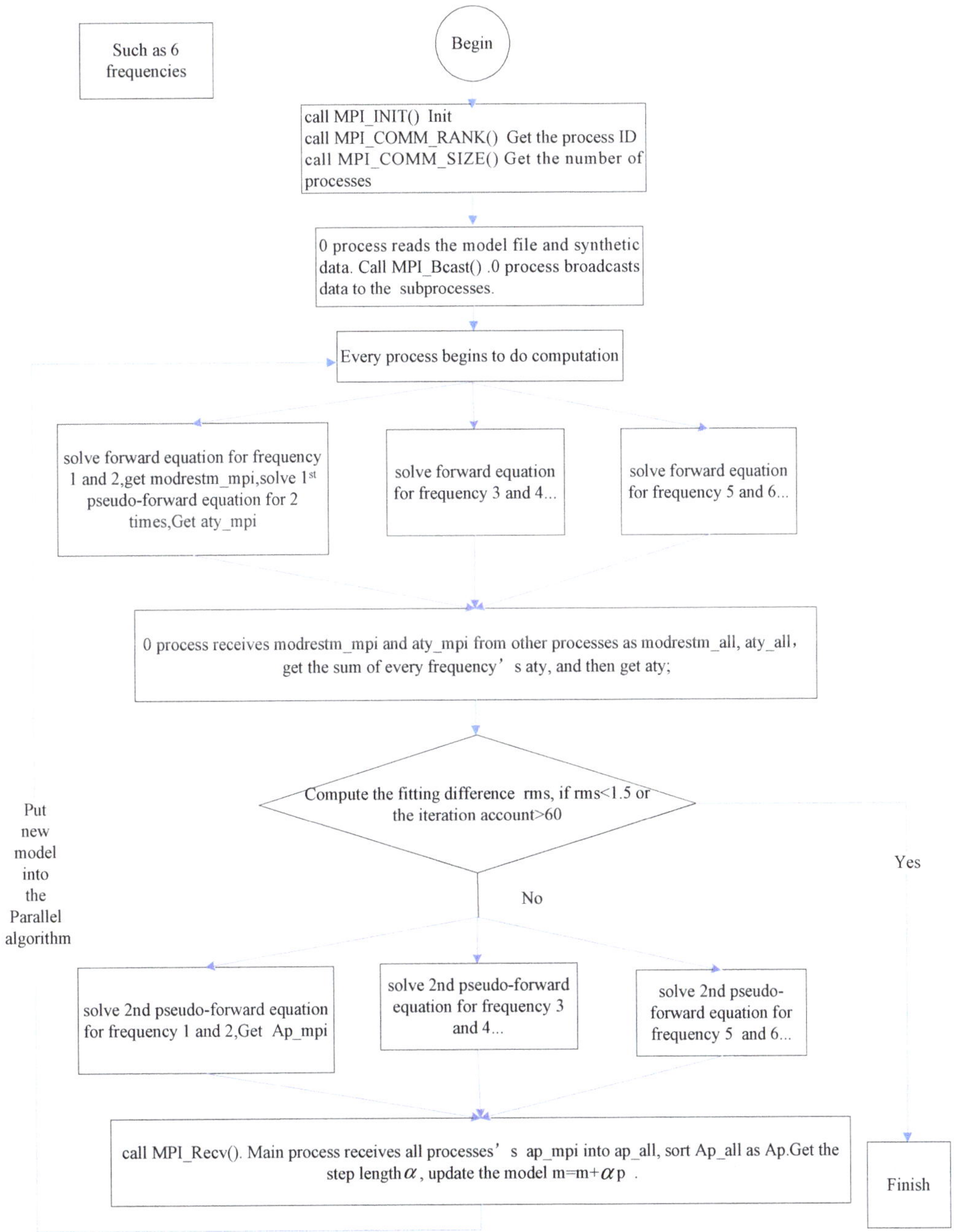

Figure 2. The flow chart of parallel computing for 2D magnetotelluric.

3.3. Computing Environment

The program is developed in the environment as shown in Table 2.

Table 2. Development environment.

OS:	Win10	Develop language:	fortran
CPU:	intel core i7 2.9 GHz support 8 processes	Compiler:	visual studio 2019
Memory:	16 GB	Parallel environment:	oneAPI HPC

4. Results

4.1. Two Low Resistivity Targets Model

The depth of the object rock is 350 m from the ground, and both of the objects are 300 m × 350 m. The width is 300 m, and the height is 350 m. The background resistivity is 100 Ω·m. The object rock's resistivity is 10 Ω·m. The distance between the two objects is 900 m. H_z data are collected in the air above the horizon. The height is 100 m. H_y data are collected on the ground. The size of the model's grid is 400 × 400, and the observed points are from point 1 to point 70 on the ground. The six frequencies range from 400 Hz to 10 Hz. We obtain the forward result including the tipper's real part and imaginary part from the parallel program.

In Figure 3, the horizontal axis is in the point position, and we choose points 12, 14, 16 ... 68, and 70 to draw the figure. When the frequency is 50 Hz, we set the result of the real part's computation to o and set the result of the real part's parallel computation to +; the result is the same. We set the result of the imaginary part's computation to ∇, and set the result of the imaginary part's parallel computation to *; the results are the same.

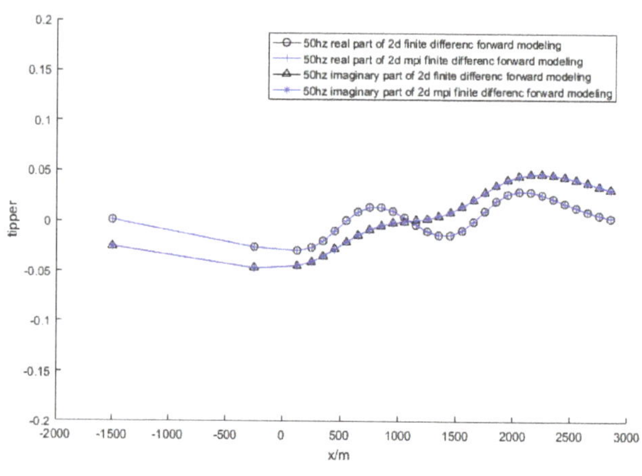

Figure 3. Forward result figure.

We set 5% of the forward data as the random error. We add the random error to the forward data and obtain the synthetic data. We take the synthetic data as the collected data and perform inversion computation. When the rms is less than 1.5, we stop the inversion and obtain the inversion result. The color bar value indicates the resistivity. The background resistivity is 100 Ω·m. The object rocks' resistivity is 10 Ω·m. We can determine that the program's result shows two object rocks' positions correctly, and the value of resistivity is approximately 10 Ω·m in the figure.

Figure 4 shows the inversion figure of parallel computation.

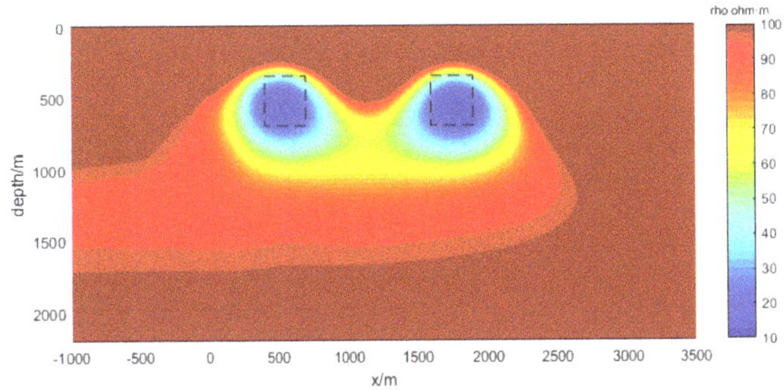

Figure 4. Inversion figure.

4.2. Proving the Validity of the Program

We analyze the serial program and find that the program calculates the equation of different frequencies separately, so the parallel algorithm distributes the work to every process and receives a result from every process. The parallel calculation result is the same as the result of the serial algorithm. The study compares the forward result of the two algorithms in Figure 3, and the data are the same. Figure 4 is the inversion result. We can observe that the experiment results show the object rocks' positions, sizes, and resistivity values correctly, so it proves the validity of the algorithm.

4.3. Parallel Efficiency

In order to evaluate the efficiency of the parallel program for different nodes, we calculate parallel speedup and parallel efficiency. Parallel speedup is the time of serial program/the time of parallel for N processes; parallel efficiency is the parallel speedup/the number of processes N. Table 3 shows the statistics of the time for the algorithm of 2D ZTEM forward modeling and inversion.

Table 3. The statistics of the time for the algorithm.

The Mode of Program	The Amount of Processes	The Amount of the Frequency Distributed for Process	The Running Time of Program (s)	Parallel Speedup	Parallel Efficiency
Serial program	1	6	9804	empty	empty
Parallel program	3	2, 2, 2	5632	1.74	58%
	6	1, 1, 1, 1, 1, 1	3069	3.19	53.2%

We can observe that the efficiency of the three processes is higher than the efficiency of the six processes in Table 2. The speedup changes and the efficiency decreases when the number of processes in the experiment is six. This is because the higher the number of processes, the more time exchanging data the processes need. Every process needs to perform computation. The computation of higher frequencies needs less time and the computation of lower frequencies needs more time. The process that performs high-frequency computation needs to wait for the process that performs the low-frequency computation. When the number of processes is six, the efficiency is lower than when the number is three. We will perform further study on the parallel algorithm.

5. Conclusions and Future Work

The computation of 2D ZTEM requires a great deal of time, so determining how to save the running time is key to this problem. We found that the parallel method can solve the problem. This study designed the parallel algorithm for 2D ZTEM forward and inversion in windows os. The experiment proved that the algorithm is correct and efficient. We will perform further study for 3D ZTEM forward and inversion based on the study.

Future work will mainly include two aspects: (1) We will optimize the algorithm with openMP and MPI mixed programming by performing an experiment, recording the running time, and calculating the efficiency of the mixed programming. Furthermore, (2) We will design the parallel algorithm for 3D ZTEM inversion.

Author Contributions: Conceptualization, H.T.; methodology, M.W. and H.T.; software, M.W.; validation, C.L. and M.P.; writing—original draft preparation, M.W.; writing—review and editing, Y.W. and M.P.; supervision, Y.W. All authors have read and agreed to the published version of the manuscript.

Funding: This study was supported by research grants from the State Key Program of National Natural Science Foundation of China (No. 41830429).

Institutional Review Board Statement: Not applicable.

Informed Consent Statement: Not applicable.

Data Availability Statement: Data sharing is not applicable.

Conflicts of Interest: The authors declare no conflict of interest. The funders had no role in the design of the study; in the collection, analyses, or interpretation of data; in the writing of the manuscript, or in the decision to publish the results.

References

1. Tan, H.; Yu, Q.; John, B.; Wei, W. Three-dimensional rapid relaxation inversion for the magnetotelluric method. *Chin. J. Geophys.* **2003**, *46*, 850–855. (In Chinese) [CrossRef]
2. Newman, G.A.; Recher, S.; Tezkan, B.; Neubauer, F.M. 3D inversion of a scalar radio magnetotelluric field data set. *Geophysics* **2003**, *68*, 791–802. [CrossRef]
3. Sasaki, Y. Three-dimensional inversion of static-shifted magnetotelluric data. *Earth Planets Space* **2004**, *56*, 239–248. [CrossRef]
4. Xu, Z.H.B. *Research of 2D ZTEM Forward Modeling and Inversion with Uneven Topography*; China University of Geosciences: Beijing, China, 2016.
5. Li, Z.Q. *Research on ZTEM 3D Forward Modeling and Inversion*; China University of Geosciences: Beijing, China, 2016.
6. Gu, G.W.; Wu, W.L.; Liang, M. The design and realization of 3D MT parallel algorithm. *Geophys. Geochem. Explor.* **2014**, *38*, 601–606.
7. Holtham, E.; Oldenburg, D.W. Three-dimensional inversion of ZTEM data. *Geophys. J. Int.* **2010**, *182*, 168–182. [CrossRef]
8. Sasaki, Y.; Yi, M.-J.; Choi, J. 3D inversion of ZTEM data from uranium exploration. *ASEG Ext. Abstr.* **2013**, *2013*, 1–4.
9. Zhdanov, M.S.; Tolstaya, E. Minimum support nonlinear parametrization in the solution of a 3D magnetotelluric inverse problem. *Inverse Probl.* **2004**, *20*, 937–952. [CrossRef]
10. Siripunvaraporn, W.; Uyeshima, M.; Egbert, G. Three-dimensional inversion for Network- Magnetotelluric data. *Earth Planets Space* **2004**, *56*, 893–902. [CrossRef]
11. Siripunvaraporn, W.; Egbert, G.; Lenbury, Y.; Uyeshima, M. Three-dimensional magnetotelluric inversion: Data-space method. *Phys. Earth Planet. Inter.* **2005**, *150*, 3–14. [CrossRef]
12. Hu, Z.Z.; Hu, X.Y. Review of three dimensional magnetotelluric inversion methods. *Prog. Geophys.* **2005**, *20*, 214–220. (In Chinese)
13. Mackie, R.L.; Madden, T.R. Three-dimensional magnetotelluric inversion using conjugate gradients. *Geophys. J. Int.* **1993**, *115*, 215–229. [CrossRef]
14. Siripunvaraporn, W. Three-Dimensional Magnetotelluric Inversion: An Introductory Guide for Developers and Users. *Surv. Geophys.* **2012**, *33*, 5–27. [CrossRef]
15. Lin, C.H.; Tan, H.D.; Tong, T. The possibility of obtaining nearby 3D resistivity structure from magnetotelluric 2D profile data using 3D inversion. *Chin. J. Geophys.* **2011**, *54*, 245–256. (In Chinese) [CrossRef]
16. Zhang, W.S.; Xue, W. *MPI on Parallel Programming Tutorial*; Qinghua University Press: Beijing, China, 2009.
17. Maris, V.; Wannamaker, P.E. Parallelizing a 3D finite difference MT inversion algorithm on a multicore PC using OpenMP. *Comput. Geosci.* **2010**, *36*, 1384–1387. [CrossRef]

applied sciences

MDPI

Article

Deep Parallel Optimizations on an LASG/IAP Climate System Ocean Model and Its Large-Scale Parallelization

Huiqun Hao [1,2], Jinrong Jiang [1,*], Tianyi Wang [3], Hailong Liu [2,4], Pengfei Lin [2,4], Ziyang Zhang [5] and Beifang Niu [1,*]

1 Computer Network Information Center, Chinese Academy of Sciences, Beijing 100190, China
2 University of Chinese Academy of Sciences, Beijing 100049, China
3 School of Information Science and Engineering, Shandong Normal University, Jinan 250014, China
4 State Key Laboratory of Numerical Modeling for Atmospheric Sciences and Geophysical Fluid Dynamics (LASG), Institute of Atmospheric Physics (IAP), Chinese Academy of Sciences, Beijing 100029, China
5 State Grid Beijing Electric Power Company, Beijing 100031, China
* Correspondence: jjr@sccas.cn; (J.J.); bniu@sccas.cn (B.N.)

Abstract: This paper proposes a series of parallel optimizations on a high-resolution ocean model, the LASG/IAP Climate System Ocean Model (LICOM), which was independently developed by the Institute of Atmospheric Physics of the Chinese Academy of Sciences. The version of LICOM that we used was LICOM 2.1. In order to improve the parallel performance of LICOM, a series of parallel optimization methods were applied. We optimized the parallelization scheme to tackle the problem of load imbalance. Some communication optimizations were implemented, including data packing, the application of the least communication algorithm, and the replacement of communications with calculations. Furthermore, for the calculation procedures, we implemented some mature optimizations and expanded functions in a loop. Additionally, a hybrid of MPI and OpenMP, as well as an asynchronous parallel IO, was used. In this work, the optimized version of LICOM 2.1 was able to achieve a speedup of more than two times compared with the original code. The parallelization scheme optimization and the communication optimization produced considerable improvement in performance in the large-scale parallelization. Meanwhile, the newly optimized LICOM could scale up to 245,760 processor cores. However, for the original version, there was no speedup when scaled up to over 10,000 processor cores. Additionally, the problem of jumpy wall time during the time integration process was also tackled with this optimization. Finally, we conducted a practical simulation from 1993 to 2007 by using the optimized version of LICOM 2.1. The results showed that the mesoscale vortex was well simulated by the model.

Keywords: LICOM; meteorological model; parallel optimization

Citation: Hao, H.; Jiang, J.; Wang, T.; Liu, H.; Lin, P.; Zhang, Z.; Niu, B. Deep Parallel Optimizations on an LASG/IAP Climate System Ocean Model and Its Large-Scale Parallelization. *Appl. Sci.* **2023**, *13*, 2690. https://doi.org/10.3390/app13042690

Academic Editors: Antonio J. Nebro and Juan A. Gómez-Pulido

Received: 31 January 2023
Revised: 13 February 2023
Accepted: 16 February 2023
Published: 19 February 2023

1. Introduction

At present, changes in the global climate and ecological environment have become some of the most important scientific problems. The ocean, as a vital part of the global climate system, has become a heated topic of research for scientists. Ocean models have considerably developed and improved since 1969, when they were developed [1]. Now, there are various kinds of global ocean models. HYCOM [2], NEMO [3], MOM [4], and LICOM [5] are some global ocean models that are used for research and production. Among them, MOM, HYCOM, and NEMO represent ocean models that were developed in Europe and the USA. LICOM was developed by scientists from IAP-CAS [6]. LICOM is widely used in the area of climate simulation and prediction, as well as the area of the numerical simulation and prediction of air–sea coupling. Different editions of LICOM were used for the ocean components of three coupled air–sea models in the Sixth Coupled

Model Intercomparison Project (CMIP6) [7,8], which included the Flexible Global Ocean–Atmosphere–Land System model version 3 [9] with a finite-volume atmospheric model (FGOALS-f3) [10], the Flexible Global Ocean–Atmosphere–Land System model version 3 with a grid-point atmospheric model (CAS FGOALS-g3) [11], and the Chinese Academy of Sciences Earth System Model (CAS-ESM) [12]. LICOM is an important tool for investigating and forecasting ocean circulation and its mechanisms. Meanwhile, as a vital part of climate system models (CSMs) and Earth system model (ESMs) [13], LICOM's performance may considerably affect climate change simulations [14,15]. As the resolution increased, the enormous computing, communication, and input/output (IO) became significant scientific and engineering challenges for scientists [16]. It will take either more time or more computing resources to run models. Widely used in many areas, including material science [17], electrochemistry [18], and meteorology, high-performance computing has become a very powerful tool in scientific research [19]. Due to the time constraints of meteorological simulation software, parallel optimization in high-performance computing is considerably important for models. Most meteorological systems suffer from the issues of low simulation speed and the inability to utilize large-scale machines. Some work on optimizing meteorological simulating systems has been done by researchers to tackle these issues. Optimizations were conducted on NEMO [20]. The overall performance was improved by 31%. The Princeton ocean model (POM) [21] was transplanted into Sunway TaihuLight [22], a very powerful supercomputer. The new edition, swPOM, was 2.8 times faster than it was on conventional supercomputers. It could be scaled up to over 250,000 cores [23]. Meanwhile, in addition to oceanic models, researchers have performed optimizations on some atmospheric models, such as IAP-AGCM [24]. The optimized code scaled up to 196,608 CPU cores, attaining a speed of 11.1 simulation years per day (SYPD) at a high resolution of 25 km [25]. In this study, in order to improve the simulation speed of LICOM, we implemented a series of optimizations and achieved a considerable speedup in comparison with the original version. The fully optimized edition of LICOM was twice as fast as the original edition. The performance experiments were conducted on the "Era", "Tianhe II", and "Tianhe III" supercomputers. Additionally, although we performed optimizations on GPUs with OpenACC [26], CUDA [27], and HIP [28], the CPU version is still widely used. For instance, many machines, such as "Tianhe III", are still pure CPU machines. Moreover, since the CPU version of LICOM is used as the ocean component of various coupled Earth system models, it is necessary to deeply optimize the CPU version.

The rest of this paper is organized as follows. The following section introduces LICOM and its control flow, along with the parallel and communication algorithm. In Section 3, the detailed optimizations that we implemented on LICOM are illustrated. Section 4 describes the experimental setups and the performance of the optimized model. Finally, Section 5 draws conclusions concerning our optimization work on LICOM.

2. The LICOM Model

LICOM is used to solve the Navier–Stokes equations. An ocean circulation model can simulate ocean temperature, salinity, velocity, and sea surface height under certain initial and boundary conditions. Meanwhile, the results of simulations can be used as the lower boundary conditions of atmospheric models and sea ice models. In addition, they can provide boundary conditions for regional ocean models. In addition, the results are capable of providing information on marine environment variables with an equably distributed space and continuous time in order to cover the shortages in the currently uneven observation data. This is beneficial for understanding the physical mechanisms of oceanographic processes. According to the original N-S equations, LICOM uses a finite-difference discrete model equation to ensure the conservation of energy and volume that are transferred during a discrete process. Since there are considerable differences in the vertical direction during the processes of marine stratification and mixing, a method with different layers is used to solve the problem. LICOM uses sea surface fluctuation with a free surface, including a surface gravity wave with a high speed and a Rossby wave with a low speed. In order to reduce the calculations, the model splits the surface wave mode and uses smaller time steps for integration, while it uses larger time steps for models that describe the vertical structure. During the process of integration, the interactions between the two time steps are kept. This method is called "the decomposition and interaction of models" [29]. The calculation of the barotropic model integral is considerably reduced through this decomposition. There are some procedures that cannot be captured by LICOM or by other ocean models, such as the procedure of turbulence. It is necessary to create processes of parameterization to describe these model-invisible procedures to realize their impacts. A low-resolution ocean model contains mesoscale vortex parameterization. However, a high-resolution ocean model (less than 10 km) can recognize mesoscale vortexes. Thus, there is no need for parameterization. Currently, LICOM has reached resolution levels of 10 km and even higher. It is capable of recognizing mesoscale vortexes and nicely simulating vortexes and their impacts. High-resolution ocean models use a corrected barotropic and baroclinic decomposition algorithm. Meanwhile, an improved double-adjustable and sticky disjunction diffusion scheme can be employed in the horizontal direction in the momentum equation and thermohaline equation. Therefore, mesoscale vortexes can be better simulated. The main control flow of LICOM is shown in Figure 1. The major processes in the integral loop include barotropic, baroclinic, and thermohaline processes, among others, and the Euler forward or leapfrog scheme is used.

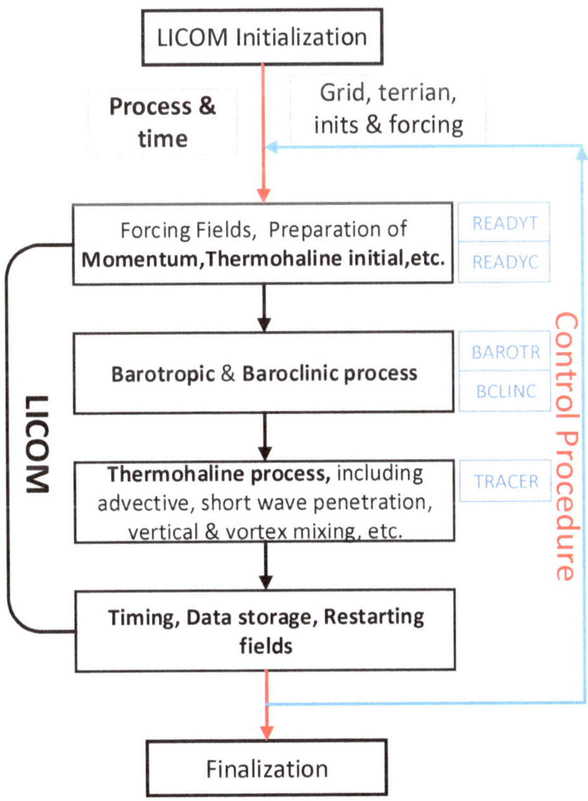

Figure 1. The control flow of LICOM.

LICOM has a variety of choices of output format, including binary files and netcdf. A binary file can be converted into a netcdf file. Thus, the results can be easily handled and investigated by using various types of professional software.

LICOM 's Parallel Scheme

The ocean is discretized into two-dimensional grid points with staggered latitude and longitude coordinates in the horizontal direction. The latitudinal direction is divided into *imt_global* grid points, and the longitudinal direction is divided into *jmt_global* grid points. The vertical direction is divided into km layers with the ocean depth as the coordinate. Thus, the ocean is decomposed into a three-dimensional structure. On this basis, the ocean grid points are organized into grid blocks. Each grid block contains $BLCKX \times BLCKY \times km$ grid points. There is barotropic-mode communication between grid blocks, and there are mutual pressure and vertical velocity calculations between layers. As shown in Figure 2, the ocean area is decomposed into $\lceil imt\_global/BLCKX \rceil \times \lceil jmt\_global/BLCKY \rceil$ grid blocks. In order to achieve a higher efficiency with the difference method, LICOM uses MPI to parallelize and speed up the calculation. *NBLOCKS_CLINIC* grid blocks are allocated to each process and sequentially calculated inside one grid block.

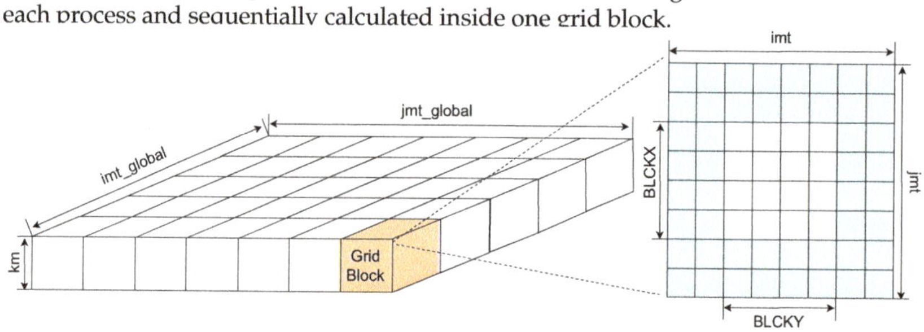

Figure 2. Ocean grid.

There is an independent array for each grid block. The size of the array decreases as the degree of parallelism increases. Since differential computation requires boundary values of adjacent grid blocks, frequent boundary communication is necessary. The array for each grid block is updated according to the array boundaries from two-dimensional logically adjacent processes. Therefore, the array of each grid block stores the boundary data of the adjacent grid block, in addition to its own data. The boundary data from other grid blocks are called the ghost boundary, while its own boundary data are called the real boundary. LICOM employs an adjustable ghost boundary strategy. The size of the ghost boundary depends on a specific issue. The blue squares in Figure 2 show the situation when the ghost equals 2. Setting the ghost to 2 instead of 1 will effectively eliminate the communication–calculation ratio, since communication is only needed for every two calculation steps.

Figure 3 illustrates the calculation procedure of a grid point in one iteration step. First, the two stripes of the ghost boundaries of each of the grid blocks are updated. The real boundary in grid blocks 1, c, and d is transferred to b and a of grid block 2. Meanwhile, the real boundary in grid blocks 2, c, and d is transferred to b and a of grid block 1. Then, the real boundaries are updated. In grid block 1, the grid points are updated from left to right until b. Meanwhile, in grid block 2, the grid points are updated from right to left until b. Finally, both grid blocks 1 and 2 update to c.

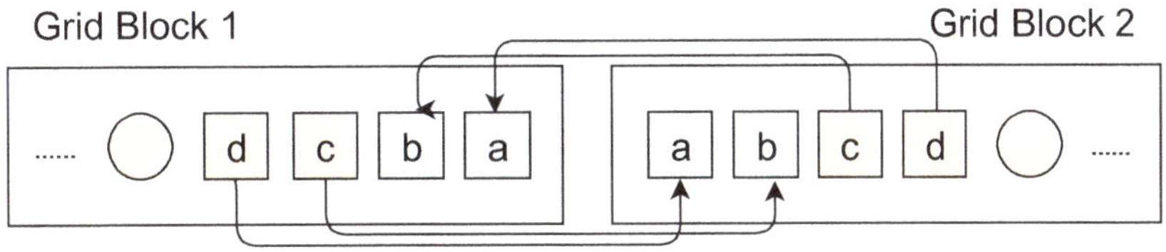

Figure 3. Ocean grid boundary communication.

3. Optimization of Parallelization

This section illustrates the series of optimizations that were implemented to speed up the running of LICOM. The original LICOM was parallelized by using only MPI. We utilized a hybrid of MPI and OpenMP in an optimized version of LICOM. Algorithm 1 shows an example of the implementation of OpenMP. However, due to the architecture of the supercomputers, the hybrid of MPI and OpenMP did not achieve good performance because the utilization of the MPI and OpenMP hybrid had no advantages at small scales. Thus, this optimization of the hybrid of MPI and OpenMP is only discussed, but is not included in the description of the performance tests in the following section. In the test on Tianhe III with a large number of PEs, we used the hybrid of MPI and OpenMP in order to utilize more PEs.

Algorithm 1: OpenMP implementation

```
OMP PARALLEL DO PRIVATE (J,I) for J←JSM to JEM do
    for I←2 to IMM do
        Calculate WKA(I,J,5) and WKA(I,J,6) ;
    end
end
```

3.1. Optimization A: Improving the Parallelization Scheme

We optimized the parallelization scheme of LICOM, as shown in Figure 4. N grid points were to be allocated to np processors. If N was divisible by np, the best decomposition was to allocate n, which equalled N divided by np, to each processor. However, in most cases, N was not divisible by np. Thus, in the original scheme, $n+1$ grid points were allocated to each processor from processor No. 0. Here, we set n to $\lfloor N/np \rfloor$. For the last processor, $N-(n+1)*(np-1)$ grid points were allocated. In the optimized scheme, n grid points were allocated to each processor, while the rest of the $N-n*np$ grid points were evenly allocated to some processors. In this way, the scalability of the model was considerably improved.

We redesigned the software structure of LICOM to improve the modernization of the software. Thus, LICOM was more convenient to use. For example, we replaced an alterable array with a fixed array, used structured data, and extracted important parameters. Therefore, there is no need to recompile or alter the code for different experiments or processor distributions.

3.2. Optimization B: Communication Optimization

In order to reduce the time consumed, we optimized the communication procedure. The method of data packing was employed to improve the communication time. In addition, the algorithms were improved by replacing communication with calculation.

(I) For a high-latitude area, a one-dimensional horizontal smoothing algorithm needs to conduct smoothing more than once. Since the times of smoothing on different latitudes are not the same, the original algorithm conducts smoothing once on each latitude, while the new algorithm involves just one smoothing and two-dimensional communication at all latitudes. Algorithms 2 and 3 show the representative code alterations. Algorithm 2 provides the original code, and Algorithm 3 describes the optimized version.

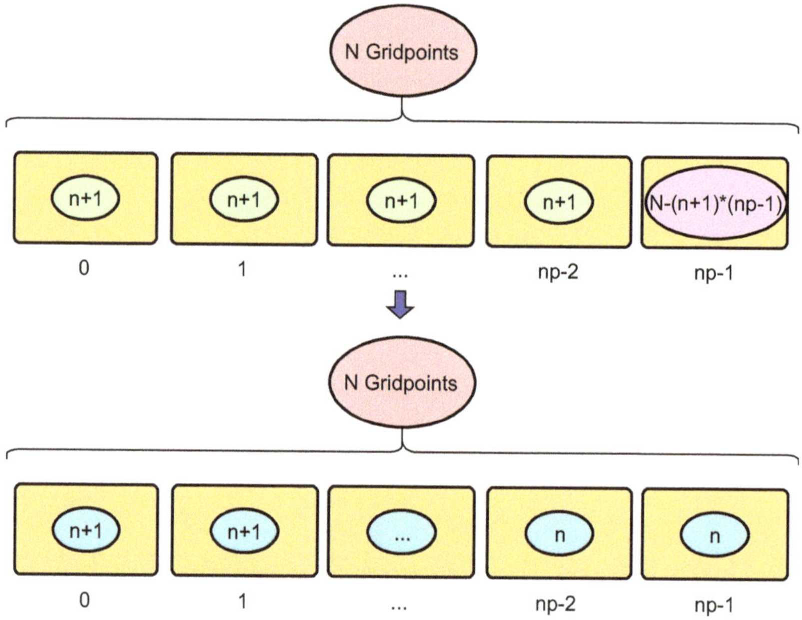

Figure 4. Optimization of the parallelization scheme.

Algorithm 2: Communication Optimization (I-A)

```
for J←JST to JMT do
    for JJ←1 to NN(j) do
        for K←1 to KK do
            for I←1 to IMT do
                Calculate XS(I);
            end
            for I←2 to IMM do
                Calculate X(I,J,K);
            end
        end
        Boundary Exchange 1D;
    end
end
```

(II) Based on previous test results when using Intel Vtune, the hotspots of LICOM are the communication functions. Therefore, the optimization of communications is of great importance. During the process of integration, the least communication algorithm was employed. Some communications were replaced by adding boundary calculations. It was beneficial to eliminate communication by adding some calculations. Because of the large number of PEs used, the total calculation job was divided by the number of PEs, while the cost of communication was multiplied by the number of PEs. Algorithms 4 and 5 show the representative code alterations. The original code in Algorithm 4 was optimized into the code in Algorithm 5 to eliminate the cost of communication.

Algorithm 3: Communication Optimization (I-B)

```
for NCY←1 to MAX_NN do
    for j←jst to jmt do
        if NN(j)≥NCY then
            for K←1 to KK do
                for I←1 to IMT do
                    Calculate XS(I);
                end
                for I←2 to IMM do
                    Calculate X(I,J,K);
                end
            end
        end
        Boundary Exchange 2D;
    end
end
```

Algorithm 4: Communication Optimization (II-A)

```
for J←JSM to JEM do
    for I←2 to IMM do
        Calculate UB(I,J), VB(I,J), and H0(I,J) ;
    end
end
Exchange boundary ub, vb, and h0 ;
for J←JSM to JEM do
    for I←2 to IMM do
        Calculate WKA(I,J,1), WKA(I,J,2), and WORK(I,J) ;
    end
end
```

Algorithm 5: Communication Optimization (II-B)

```
for J←JST to JET do
    for I←1 to IMT do
        Calculate UB(I,J), VB(I,J), and H0(I,J) ;
    end
end
for J←JST to JET do
    for I←1 to IMT do
        Calculate WKA(I,J,1), WKA(I,J,2), and WORK(I,J) ;
    end
end
```

(III) The grid matching information was obtained via communication in the original code in Listing 1. In contrast, in the optimized algorithm, it was obtained by conducting a calculation in Listing 2. Therefore, point-to-point communication was reduced by a considerable amount.

Listing 1. Communication Optimization (III-A)

```
if (mytid == 0) then
  i_start(1)= i_global(1)
  j_start(1)= j_global(1)
  do n=1,nproc-1
    call mpi_recv(j_start(n+1),1,mpi_integer,n,tag_1d,&
                  mpi_comm_ocn,status,ierr)
    call mpi_recv(i_start(n+1),1,mpi_integer,n,tag_2d,&
                  mpi_comm_ocn,status,ierr)
  end do
else
  j_start(1) =j_global(1)
  i_start(1) =i_global(1)
  call mpi_send(j_start(1),1,mpi_integer,0,tag_1d,&
                mpi_comm_ocn,ierr)
  call mpi_send(i_start(1),1,mpi_integer,0,tag_2d,&
                mpi_comm_ocn,ierr)
end if
```

Listing 2. Communication Optimization (III-B)

```
do i=1,nproc
  iix=mod(i-1,nx_proc)
  iiy=(i-1-iix)/nx_proc
  i_start(i)=iix*(imt-num_overlap)+1
  j_start(i)=3
  if(iiy/=0) then
    do j=1,iiy
      j_start(i)=j_start(i)+jmt-num_overlap
    end do
  endif
end do
```

3.3. Optimization C: Calculation Optimization

We carried out a series of mature optimizing methods, including vectorization, memory access optimization, instruction optimization, cache optimization, and runtime optimization. Thus, we could fully discover the potential performance of stencil calculation on the CPU cluster, such as by expanding the functions in a loop, as shown in Algorithms 6 and 7. The original DENS function in Algorithm 6 was expanded in the optimized version in Algorithm 7 in order to utilize vectorization.

Algorithm 6: Function Expansion A

```
for K←1 to KMM1 do
    for J←JST to JET do
        for I←2 to IMM do
            Calculate TUP, SUP, and RHOUP ;
            Calling DENS() function during calculation of RHOUP ;
        end
    end
end
```

Algorithm 7: Function Expansion B

```
for K←1 to KMM1 do
    for J←JST to JET do
        for I←2 to IMM do
            Calculate TUP, SUP, and RHOUP ;
            Instead of calling the DENS() function, the calculation procedure is
              written out in detail ;
        end
    end
end
```

3.4. Optimization D: Parallel IO

We also designed an asynchronous parallel IO method. As shown in Figure 5, an independent node was employed to conduct the IO procedures. A separate communication group was created for the IO work. When the computing nodes needed to carry out IO procedures after calculation procedures, the communication group of the computing nodes collected the data and sent it to the IO node. After receiving all of the data, the IO node began IO procedures. Meanwhile, the nodes from the computing group could continue with their calculation procedures instead of waiting for the completion of the IO procedures. Therefore, the elapsed time for IO was hidden by overlapping the calculations and IO.

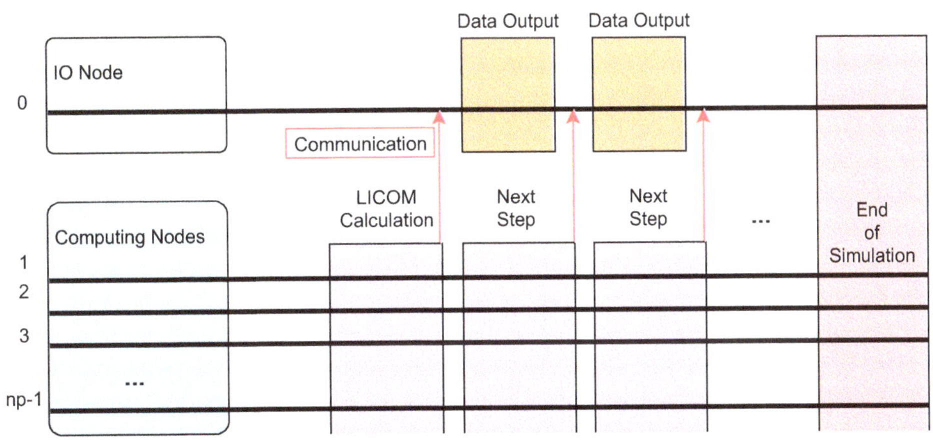

Figure 5. Asynchronous parallel IO.

4. Parallel Performance and Application in Actual Scenarios

Overall, we ran three sets of tests. The setup of the tests is shown in Table 1.

Table 1. Setup of Tests.

No.	Description	Machine
1	Performance test with three versions of code	Era and Tianhe II
2	Scalability test	Tianhe III
3	Real scenario test	Era

For the first test, we used three versions of the code for comparison. The code versions were (a) the original LICOM, (b) semi-optimized LICOM, in which optimization methods B(I), B(III), and C were employed, and (c) fully optimized LICOM. The hardware environments were the "Era" supercomputer and the "Tianhe II" supercomputer. For the second test, the fully optimized code was run on the "Tianhe III" supercomputer. For the third test, the fully optimized code was run on the "Era" supercomputer for a long term in a real scenario. For all of the tests on the three different hardware platforms, the same test case with a global resolution of 10 km × 10 km was used. The simulation started from 1993. The detailed configurations are listed in Table 2. These parameters controlled the run of LICOM and described the real scenario in our test. An elapsed time of one model day was used as an indicator for comparison. We ran each particular test at least five times to calculate the average value. The simulation year per day, as a widely used indicator of running speed, was calculated from the elapsed time of one model day.

Table 2. Configurations of the test cases.

Module	LICOM2.1
Horizontal resolution	0.1° (lat) × 0.1°(lon)
Vertical levels	55 levels
Grid point	3600 × 1683
Advection scheme	Shape-preserving [30]
Vertical mixing	Canuto [31]
Mesoscale eddy	Gent and McWilliams [32]
Horizontal viscosity	3×10^3 m^2s^{-1}
Forcing formula and dataset	Large and Yeager [33]; COREs

4.1. *Performance on Era*

The CPU on the computing nodes of Era was an Intel(R) CPU E5-2680V3:2.5GHz. The operating system was Linux CentOS release 6.4 (Final). The compilers were Intel composer_xe_2013_sp1.0.080 and Intelmpi 4.1.3.049. Based on the above environment, we conducted tests using up to 4800 processor cores. Since LICOM is so complicated, the elapsed time of LICOM when simulating one model day is used as an indicator of performance. Figure 6 shows the elapsed time and running speed of the three editions of LICOM, while Figure 7 shows the speedups. The simulation year per day (SYPD) is usually used to measure the computational performance of models. The speedups for the three editions of code were calculated separately. The elapsed times for each edition of code on 1200 PEs were chosen as references. For instance, the speedup of the original code on 4800 PEs over 1200 PEs was the elapsed time of the original code on 1200 PEs divided by the elapsed time of the original code on 4800 PEs. The speedup of the fully optimized code on 4800 PEs over 1200 PEs was the elapsed time of the fully optimized code on 1200 PEs divided by the elapsed time of the fully optimized code on 4800 PEs. The reason for why the elapsed times of 1200 PEs were chosen as references was that, with a resolution of 10 km × 10 km, LICOM needed a very large memory space. Thus, we needed more PEs so that after decomposition, the limited memory space on each PE could meet the demands of LICOM. As shown in Figures 6 and 7, the semi-optimized LICOM was much faster than the original LICOM, but it still suffered from the problem of scalability. However, the fully optimized LICOM with the optimization of the decomposition scheme showed considerably good scalability when 4800 processor cores were used. The computing speed reached 9 model years per day. Additionally, the elapsed times of both the original and semi-optimized code on 3600 PEs were longer than those on 2400 PEs. This was a non-intuitive phenomenon. However, for the fully optimized code, there was no similar non-intuitive phenomenon. We can infer that the problem that caused the non-intuitive phenomenon was tackled by the optimizations in the fully optimized code. Therefore, the possible reasons that caused this non-intuitive phenomenon could have been the communication overhead and load imbalance.

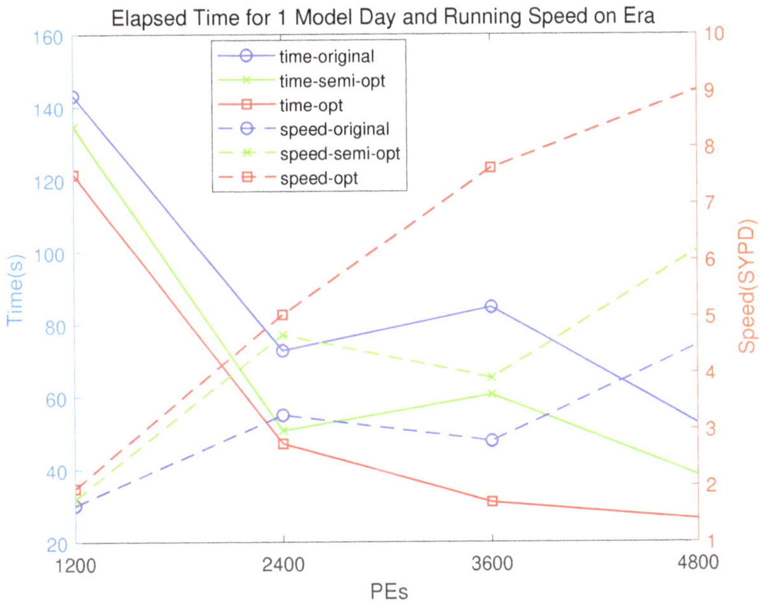

Figure 6. Elapsed time and speed of LICOM on Era.

4.2. Performance on Tianhe II

Based on the test results on Era, we decided to conduct tests on a larger scale. The hardware environment was Tianhe II. The processor was an Intel Ivy Bridge-E Xeon E5-2692V2:2.2GHz. The operating system edition was Red Hat Enterprise Linux Server release 6.5 (Santiago). The compilers were Intel composer_xe_2013_sp1.2.144 and MPICH 3.1.3. In the above hardware environment, we conducted tests on the same three code editions. As shown in Figures 8 and 9, the fully optimized LICOM achieves a good speedup when 9600 and 19,200 processor cores were used. The computing speed reached 12.6 model years per day, which was twice the speed of the original LICOM. However, as shown in Figure 9, a non-intuitive phenomenon occurred. The speedup for the semi-optimized code on 9600 PEs was smaller than that on 4800 PEs. Similarly to the phenomenon on Era, this may have been due to the communication overhead and load imbalance. In contrast, the fully optimized code achieved good speedups on 9600 and 19,200 PEs. This showed that our optimization worked well in improving the scalability of the code.

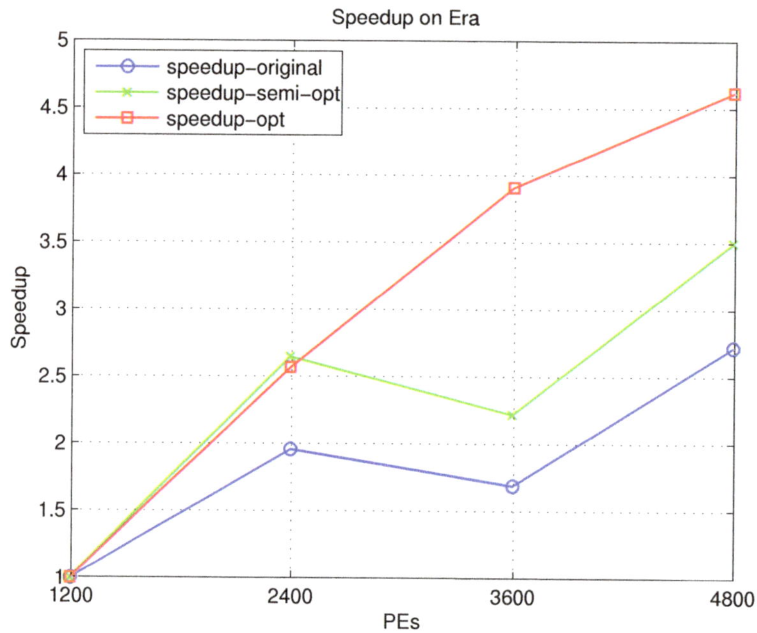

Figure 7. Speedup of LICOM on Era.

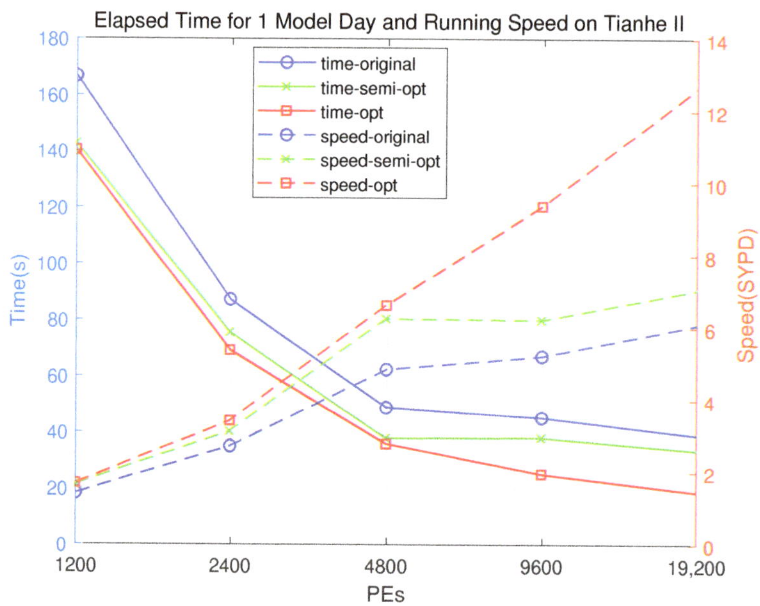

Figure 8. Elapsed time and speed of LICOM on Tianhe II.

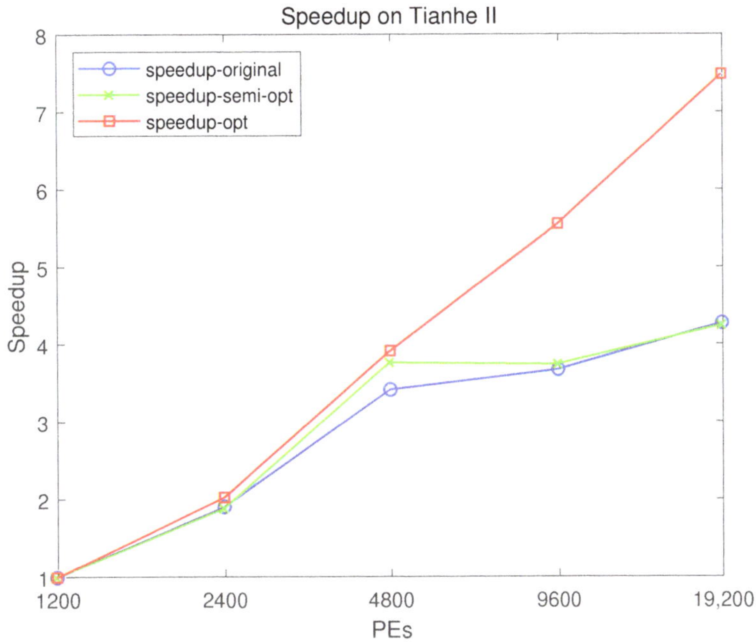

Figure 9. Speedup of LICOM on Tianhe II.

4.3. Performance on Tianhe III

We tested the fully optimized LICOM on the prototype system of the Tianhe III supercomputer. We reached up to 245,760 PEs, which was the summation of the CPU cores and the Matrix-2000 cores. Figure 10 shows the performance of LICOM on various numbers of PEs. The speedup in Figure 11 is the running time on each PE count divided by the running time on 1920 PEs. The reason for why 1920 PEs were chosen as the reference was that the elapsed time on 960 PEs was too long, which may lead to an abnormal speedup diagram. The speedup diagram in Figure 11 shows that there is still potential for optimization, since the speedup fell when more than 61,400 cores were used.

For the speedups on the three supercomputers, we can see that on Era and Tianhe II, there were still good speedups when 4800 and 19,200 PEs were used. The reason for why we did not test on more PEs was that we did not get access due to the policy for the supercomputer. However, on Tianhe III, the speedup was lower when more PEs were used. This might have been due to the communication overhead. Additionally, on a similar number of PEs, Era achieved the best speedup. For instance, on 4800 PEs (7680 on Tianhe III), the speedup for the fully optimized code on the three platforms was about 4.6, 3.9, and 3, respectively. A possible reason might be that Era had the CPU with the best performance.

4.4. Experiment on the Application of LICOM in a Real Scenario

Moreover, we conducted a set of application tests. LICOM was used with 10 km × 10 km as the resolution. CORE-II was employed as a forcing field. The simulation period was from 1993 to 2007. The output included temperature, salinity, sea surface height, and the 3D current field. Figure 12 shows the abnormal sea surface height value (difference from the average value) on 31 December 2007. As can be seen, an eddy was apparent. Additionally, there was a special difference. The middle image in Figure 12 shows the average number of eddies on every grid point. West and east boundary eddies often occurred in the southern ocean. There were at least 50 eddies that occurred in some areas. Therewasis a considerable difference between the structures of the eddies of anticyclones and cyclones, as shown in the bottom image in Figure 12. An anticyclone eddy had a sunken center with a temperature increment of 2°, while a cyclone eddy had a raised center with a temperature decline of 2°.

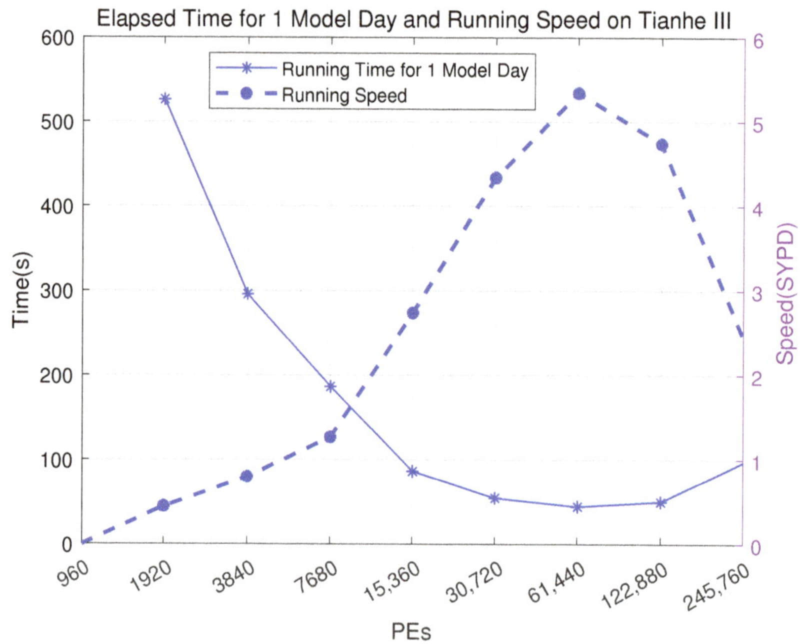

Figure 10. Elapsed time and speed of LICOM on Tianhe III.

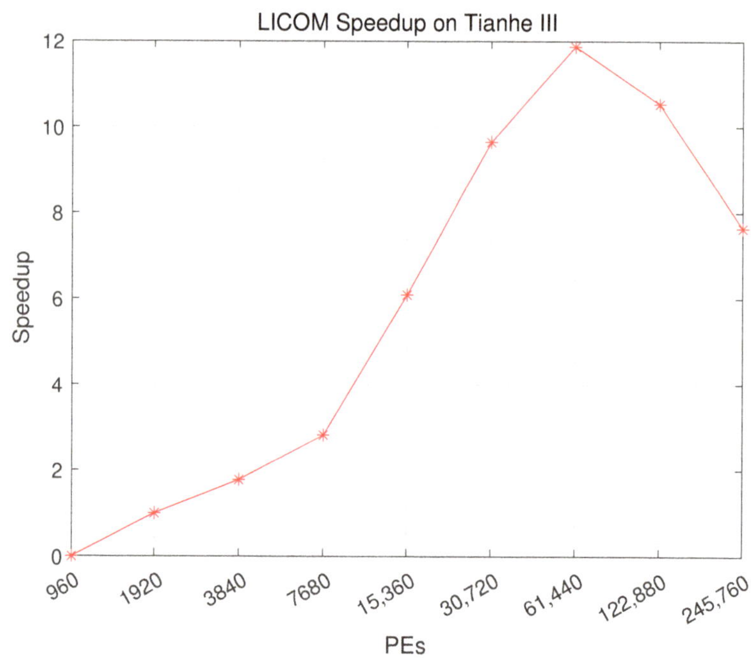

Figure 11. Speedup of LICOM on Tianhe III.

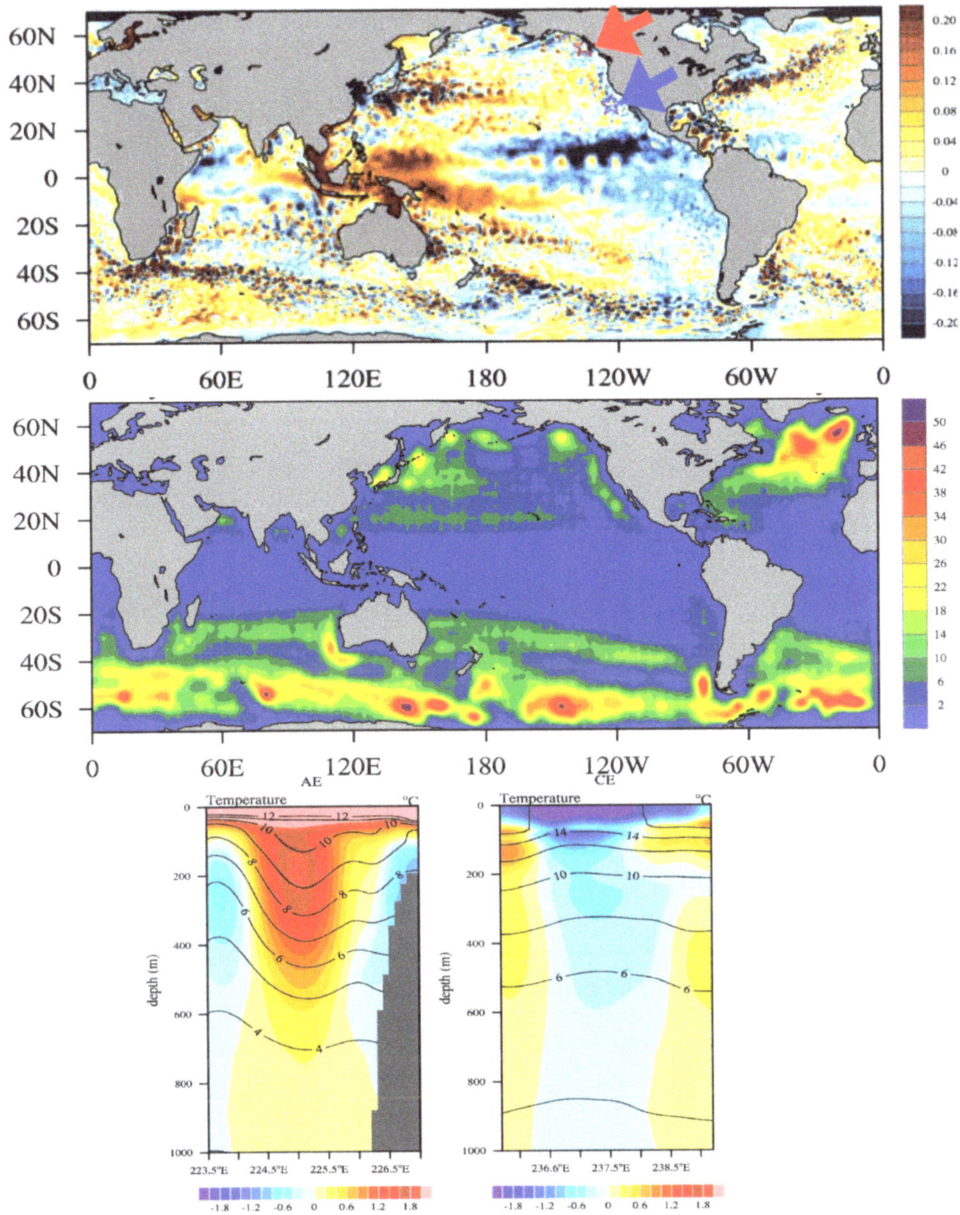

Figure 12. Output of the LICOM application.

In addition, we compared the results produced by the original code and the fully optimized code to see whether there were any differences. Figure 13 shows the results of the sea surface temperature and the differences between the two editions.

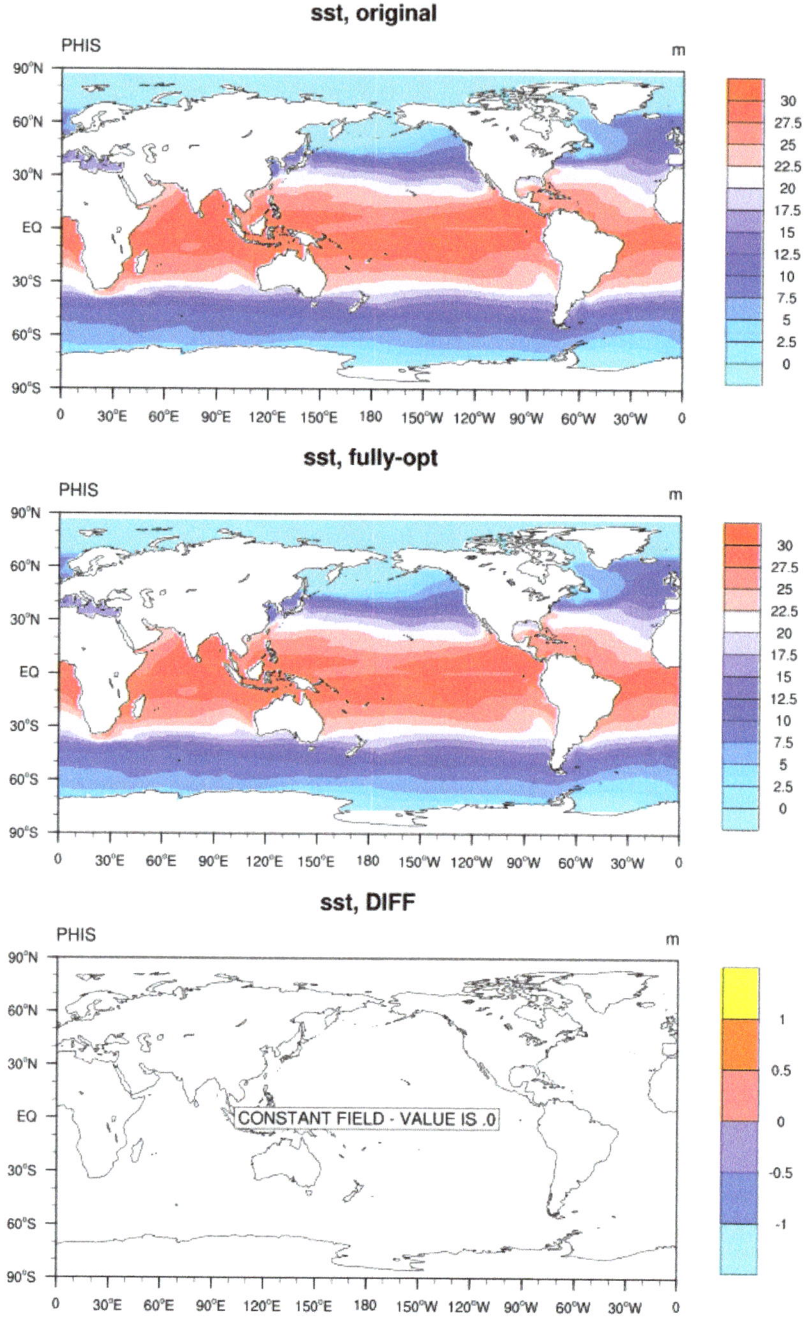

Figure 13. Correctness of simulations.

5. Conclusions

High-resolution general ocean circulation models are also called "eddy-resolving" ocean circulation models, and they are usually models with resolutions higher than 10 km. This kind of model is capable of simulating the characteristics of mesoscale eddies and their climate effects. Moreover, an "eddy-resolving" ocean model can describe submarine topography, the land–sea distribution, and the spatial and temporal structure of an ocean's west boundary circulation well. Therefore, the development of global "eddy-resolving" ocean circulation models has been drawing people's attention in the field of physical oceanography and climate research.

In this work, we applied several parallel optimization methods on LICOM, including improvements in the parallelization scheme, communication optimization, the floating-point performance, the asynchronous IO, hybrid programming of MPI and OpenMP, and the redesign of the software structure. The performance of the distributed cluster was fully utilized. The computing speed of the optimized version of LICOM reached 12.6 model years per day when 19,200 processor cores were used, which was twice that of the original LICOM. The optimized LICOM could scale up to 245,760 processor cores. However, for the old version, there would not be much of a speedup when more than 19,200 processor cores were used. This is a vital improvement thanks to the optimization in this work. As mentioned in Section 1, swPOM can be scaled up to 250,000 cores. Although it is not appropriate to simply compare the scalability of different systems on different machines, the results of our work are around the same level as that of other researchers' work. We found that the optimization of communications and the tackling of load imbalance have considerable benefits in improving the performance of LICOM according to our test results.

In addition, we conducted simulations of a real scenario from 1993 to 2007 by using the optimized LICOM. The results showed that mesoscale vortexes were well simulated by the model. In conclusion, our optimization work considerably improved the performance of LICOM in terms of computing speed and scalability.

Author Contributions: Conceptualization, H.L., P.L., B.N. and J.J. ; methodology, J.J.; software, T.W., H.H. and J.J.; validation, H.H. and T.W.; formal analysis, B.N.; investigation, H.H.; resources, J.J.; data curation, H.H. and T.W.; writing—original draft preparation, H.H.; writing—review and editing, J.J.; visualization, P.L. and Z.Z.; supervision, J.J.; project administration, J.J.; funding acquisition, J.J. All authors have read and agreed to the published version of the manuscript.

Funding: This research was supported by the National Natural Science Foundation of China (Grant No. 41931183) and the National Key Scientific and Technological Infrastructure project, "Earth System Science Numerical Simulator Facility" (EarthLab).

Institutional Review Board Statement: Not applicable.

Informed Consent Statement: Not applicable.

Data Availability Statement: Data sharing is not applicable.

Conflicts of Interest: The authors declare no conflict of interest. The funders had no role in the design of the study; in the collection, analyses, or interpretation of data; in the writing of the manuscript; or in the decision to publish the results.

Abbreviations

The following abbreviations are used in this manuscript:

IO	Input and Output
LICOM	LASG/IAP Climate System Ocean Model
LASG	State Key Laboratory of Numerical Modeling for Atmospheric Sciences Geophysical Fluid Dynamics
IAP	Institute of Atmospheric Physics
HYCOM	Hybrid Coordinate Ocean Model
NEMO	Nucleus for the European Modeling of the Ocean
MOM	Modular Ocean Model
CMIP6	The Sixth Coupled Model Intercomparison Project

FGOALS-f3	Flexible Global Ocean–Atmosphere–Land System model version 3 with a finite-volume atmospheric model
FGOALS-g3	Flexible Global Ocean–Atmosphere–Land System model version 3 with a grid-point atmospheric model
CAS-ESM	Chinese Academy of Sciences Earth System Model
CSM	Climate System Model
ESM	Earth System Model
CPU	Central Processing Unit
CUDA	Compute Unified Device Architecture
HIP	Heterogeneous Compute Interface for Portability

References

1. Zhang, X.; Yu, Y.; Liu, H. The development and application of ocean circulation model I. Global general ocean circulation model. *Chin. J. Atmos. Sci.* **2003**, *27*, 607–617.
2. Wallcraft, A.J.; Metzger, E.J.; Carroll, S.N. Software Design Description for the HYbrid Coordinate Ocean Model (HYCOM), version 2.2. 2009. Available online: https://apps.dtic.mil/sti/citations/ADA494779 (accessed on 10 August 2022)
3. Madec, G.; Bourdallé-Badie, R.; Bouttier, P.A.; Bricaud, C.; Bruciaferri, D.; Calvert, D.; Chanut, J.; Clementi, E.; Coward, A.; Delrosso, D.; et al. NEMO ocean engine. In *Scientific Notes of IPSL Climate Modelling Center (v4.2, Number 27)*; Zenodo: Honolulu, HW, USA, 2022. [CrossRef]
4. Pacanowski, R.C.; Dixon, K.W.; Rosati, A. GFDL Modular Ocean Model, Users Guide Version 1.0. *Gfdl. Tech. Rep.* **1991**, *2*, 142.
5. Liu, H.; Lin, P.; Yu, Y.; Zhang, X. The Baseline Evaluation of LASG/IAP Climate System Ocean Model (LICOM) Version 2. *Acta Meteorol. Sin.* **2012**, *26*, 318–329. [CrossRef]
6. Liu, H.; Yu, Y.; Liu, X.; Zhang, X. The Development of LASG/IAP Climate System Ocean Circulation Model(LICOM)(Abstract). In Proceedings of the Chinese Meteorological Society Annual Conference, Beijing, China, 8 December 2003.
7. Lin, P.; Yu, Z.; Liu, H.; Yu, Y.; Li, Y.; Jiang, J.; Xue, W.; Chen, K.; Yang, Q.; Zhao, B.; et al. LICOM Model Datasets for the CMIP6 Ocean Model Intercomparison Project. *Adv. Atmos. Sci.* **2020**, *37*, 239–249. [CrossRef]
8. Griffies, S.; Danabasoglu, G.; Durack, P.; Adcroft, A.; Balaji, V.; Böning, C.; Chassignet, E.; Curchitser, E.; Deshayes, J.; Drange, H.; et al. OMIP contribution to CMIP6: Experimental and diagnostic protocol for the physical component of the Ocean Model Intercomparison Project. *Geosci. Model Dev.* **2016**, *9*, 3231–3296. [CrossRef]
9. Li, Y.W.; Liu, H.L.; Ding, M.R.; Lin, P.F.; Yu, Z.P.; Meng, Y.; Li, Y.L.; Jian, X.; Jiang, J.; Chen, K.; et al. Eddy-resolving Simulation of CAS-LICOM3 for Phase 2 of the Ocean Model Intercomparison Project. *Adv. Atmos. Sci.* **2020**, *37*, 1067–1080. 10.1007/s00376-020-0057-z. [CrossRef]
10. He, B.; Yu, Y.; Bao, Q.; Lin, P.F.; Liu, H.L.; Li, J.X.; Wang, L.; Liu, Y.M.; Wu, G.; Chen, K.; et al. CAS FGOALS-f3-L model dataset descriptions for CMIP6 DECK experiments. *Atmos. Ocean. Sci. Lett.* **2020**, *13*, 582–588. [CrossRef]
11. Li, L.; Yu, Y.; Tang, Y.; Lin, P.; Xie, J.; Song, M.; Dong, L.; Zhou, T.; Liu, L.; Wang, L.; et al. The Flexible Global Ocean Atmosphere Land System Model Grid Point Version 3 (FGOALS-g3): Description and Evaluation. *J. Adv. Model. Earth Syst.* **2020**, *12*, 9. [CrossRef]
12. Zhang, H.; Zhang, M.; Jin, J.; Fei, K.; Ji, D.; Wu, C.; Zhu, J.; He, J.; Chai, Z.; Xie, J.; et al. CAS-ESM 2: Description and climate simulation performance of the Chinese Academy of Sciences (CAS) Earth System Model (ESM) version 2. *J. Adv. Model. Earth Syst.* **2020**, *12*, 12. [CrossRef]
13. Craig, A.P.; Vertenstein, M.; Jacob, R. A new flexible coupler for earth system modeling developed for CCSM4 and CESM1. *Int. J. High Perform. Comput. Appl.* **2012**, *26*, 31–42. [CrossRef]
14. Jiang, J.; Wang, T.; Chi, X.; Hao, H.; Wang, Y.; Chen, Y.; Zhang, H. SC-ESAP: A Parallel Application Platform for Earth System Model. *Procedia Comput. Sci.* **2016**, *80*, 1612–1623. [CrossRef]
15. Liu, H.; Lin, P.; Zheng, W.; Luan, Y.; Ma, J.; Ding, M.; Mo, H.; Wan, L.; Ling, T. A global eddy-resolving ocean forecast system in China—LICOM forecast system (LFS). *J. Oper. Oceanogr.* **2021**, 1–13. [CrossRef]
16. Palmer, T. Climate forecasting: Build high-resolution global climate models. *Nat. News* **2014**, *515*, 338–339. 10.1038/515338a. [CrossRef]
17. Bahadur, A.; Iqbal, S.; Shoaib, M.; Saeed, A. Electrochemical study of specially designed graphene-Fe3O4-polyaniline nanocomposite as a high-performance anode for lithium-ion battery. *Dalton Trans. Int. J. Inorg. Chem.* **2018**, *47*, 15031–15037. [CrossRef]
18. Ditta, N.A.; Yaqub, M.; Nadeem, S.; Jamil, S.; Hassan, S.U.; Iqbal, S.; Javed, M.; Elkaeed, E.B.; Alshammari, F.H.; Alwadai, N. Electrochemical Studies of LbL Films With Dawson Type Heteropolyanion Glassy Carbon Electrode Sensor Modified for Methyl Parathion Detection. *Front. Mater.* **2022**, *9*, 877683. [CrossRef]
19. Chi, X.; Hu, Y. The Current Supercomputing Development of China. *Res. World* **2013**, *8*, 56–60.
20. Zhou, S.; Liu, W.; Song, Z.; Yang, X. Code modernization optimization of ocean general circulation model NEMO. *Adv. Mar. Sci.* **2021**, *39*, 62–67. [CrossRef]
21. Mellor, G.L. *User's Guide for a Three Dimentional, Primitive Equation, Numerical Ocean Model*; Program in Atmospheric and Oceanic Sciences: Princeton, NJ, USA, 1998.

22. Fu, H.; Liao, J.; Yang, J.; Wang, L.; Song, Z.; Huang, X.; Yang, C.; Xue, W.; Liu, F.; Qiao, F.; et al. The Sunway TaihuLight supercomputer: System and applications. *Sci. China Inf. Sci.* **2016**, *59*, 072001. [CrossRef]
23. Wu, Q.; Ni, Y.; Huang, X. Regional Ocean Model Parallel Optimization in "Sunway TaihuLight". *J. Comput. Res. Dev.* **2019**, *56*, 1556–1566. [CrossRef]
24. Zhang, H.; Lin, Z.; Zeng, Q. The computational scheme and the test for dynamical framework of IAP AGCM-4. *Chinese J. Atmos. Sci.* **2009**, *33*, 1267–1285.
25. Cao, H.; Yuan, L.; Zhang, H.; Zhang, Y.; Wu, B.; Li, K.; Li, S.; Zhang, M.; Lu, P.; Xiao, J. AGCM-3DLF: Accelerating Atmospheric General Circulation Model via 3D Parallelization and Leap-Format. *Distrib. Parallel Clust. Comput.* **2021**, *14*, 8. [CrossRef]
26. Jiang, J.; Lin, P.; Wang, J.; Liu, H.; Chi, X.; Hao, H.; Wang, Y.; Wang, W.; Zhang, L. Porting LASG/IAP Climate System Ocean Model to Gpus Using OpenAcc. *IEEE Access* **2019**, *7*, 154490–154501. [CrossRef]
27. Wei, J.; Jiang, J.; Liu, H.; Zhang, F.; Lin, P.; Wang, P.; Yu, Y.; Chi, X.; Zhao, L.; Ding, M. LICOM3-CUDA: A GPU version of LASG/IAP climate system ocean model version 3 based on CUDA. *J. Supercomput.* **2023**, 1–31. [CrossRef]
28. Wang, P.; Jiang, J.; Lin, P.; Ding, M.; Wei, J.; Zhang, F.; Zhao, L.; Li, Y.; Yu, Z.; Zheng, W.; et al. The GPU version of LASG/IAP Climate System Ocean Model version 3 (LICOM3) under the heterogeneous-compute interface for portability (HIP) framework and its large-scale application. *Geosci. Model Dev.* **2021**, *14*, 2781–2799. [CrossRef]
29. Blumberg, A.F.; Mellor, G.L. Three-Dimensional Coastal Ocean Models. *Coast. Estuar. Sci.* **1987**, *32*, 1–16.
30. Yu, R. A two-step shape-preserving advection scheme. *Adv. Atmos. Sci.* **1994**, *11*, 479–490.
31. Canuto, V.M.; Howard, A.; Cheng, Y.; Dubovikov, M.S. Ocean turbulence. Part I: One-point closure modelmomentum and heat vertical diffusivities. *J. Phys. Oceanogr.* **2001**, *31*, 1413–1426. [CrossRef]
32. Gent, P.; Mcwilliams, J.C. Isopycnal mixing in ocean circulation models. *J. Phys. Oceanogr.* **1990**, *20*, 150–155. [CrossRef]
33. Large, W.; Yeager, S. *Diurnal to Decadal Global Forcing for Ocean and Sea-Ice Models: The Data Sets and Flux Climatologies*; NCAR/TN-460+STR; OpenSky Press: Austin, TX, USA, 2004; pp. 1–105.

applied sciences

MDPI

Article

Spatial–Temporal Correlation Considering Environmental Factor Fusion for Estimating Gross Primary Productivity in Tibetan Grasslands

Qinmeng Yang [1], Ningming Nie [1,2], Yangang Wang [1,2,*], Xiaojing Wu [3,4], Weihua Liu [2,3,4], Xiaoli Ren [3,4], Zijian Wang [1], Meng Wan [1] and Rongqiang Cao [1,2]

[1] Computer Network Information Center, Chinese Academy of Sciences, Beijing 100190, China; qmyang@cnic.cn (Q.Y.); nienm@sccas.cn (N.N.); wangzj@cnic.cn (Z.W.); wanmengdamon@cnic.cn (M.W.); caorq@sccas.cn (R.C.)
[2] University of Chinese Academy of Sciences, Beijing 100049, China; liuwh.20b@igsnrr.ac.cn
[3] Key Laboratory of Ecosystem Network Observation and Modeling, Institute of Geographic Sciences and Natural Resources Research, Chinese Academy of Sciences, Beijing 100101, China; wuxj@igsnrr.ac.cn (X.W.); renxl@igsnrr.ac.cn (X.R.)
[4] National Ecosystem Science Data Center, Institute of Geographic Sciences and Natural Resources Research, Chinese Academy of Sciences, Beijing 100101, China
* Correspondence: wangyg@sccas.cn; Tel.: +86-13718268975

Abstract: Gross primary productivity (GPP) is an important indicator in research on carbon cycling in terrestrial ecosystems. High-accuracy GPP prediction is crucial for ecosystem health and climate change assessments. We developed a site-level GPP prediction method based on the GeoMAN model, which was able to extract spatiotemporal features and fuse external environmental factors to predict GPP on the Tibetan Plateau. We evaluated four models' behavior—Random Forest (RF), Support Vector Machine (SVM), Deep Belief Network (DBN), and GeoMAN—in predicting GPP at nine flux observation sites on the Tibetan Plateau. The GeoMAN model achieved the best results (R^2 = 0.870, RMSE = 0.788 g Cm^{-2} d^{-1}, MAE = 0.440 g Cm^{-2} d^{-1}). Distance and vegetation type of the flux sites influenced GPP prediction, with the latter being more significant. The different grassland vegetation types exhibited different sensitivity to environmental factors (Ta, PAR, EVI, NDVI, and LSWI) for GPP prediction. Among them, the site located in the alpine swamp meadow was insensitive to changes in environmental factors; the GPP prediction accuracy of the site located in the alpine meadow steppe decreased significantly with the changes in environmental factors; and the GPP prediction accuracy of the site located in the alpine Kobresia meadow also varied with environmental factor changes, but to a lesser extent than the former. This study provides a good reference that deep learning model is able to achieve good accuracy in GPP simulation when considers spatial, temporal, and environmental factors, and the judgement made by deep learning model conforms to basic knowledge in the relevant field.

Keywords: deep learning; GeoMAN model; gross primary productivity; attention mechanism; interdisciplinary

Citation: Yang, Q.; Nie, N.; Wang, Y.; Wu, X.; Liu, W.; Ren, X.; Wang, Z.; Wan, M.; Cao, R. Spatial–Temporal Correlation Considering Environmental Factor Fusion for Estimating Gross Primary Productivity in Tibetan Grasslands. *Appl. Sci.* **2023**, *13*, 6290. https://doi.org/10.3390/app13106290

Academic Editor: Nathan J Moore

Received: 25 April 2023
Revised: 14 May 2023
Accepted: 18 May 2023
Published: 21 May 2023

1. Introduction

Gross primary productivity (GPP) is the cumulative sum of organisms produced by plants absorbing CO_2 during photosynthesis [1,2]. It drives the seasonal and annual variations in atmospheric CO_2 concentration, which reflect the production capacity of terrestrial ecosystems under natural conditions [2,3], and is an important indicator for assessing ecosystem health and climate change [4]. Therefore, accurate prediction of GPP is crucial for ecosystem function evaluation and carbon balance research [2].

The common methods for quantifying and predicting GPP are based on processing observational data and process-based model simulation [2,5]. Observational data include data obtained using the eddy covariance (EC) technique, in which GPP values are obtained by calculating net ecosystem exchange (NEE) from vertical turbulent transport in the atmosphere under meteorological conditions [5,6], and satellite data, which are commonly used for GPP estimation due to their stability and sustainability, such as the MODIS GPP standard product, the VPM model, and the EC-LUE model [2,7–9]. However, satellite GPP products cannot fully guarantee the reliability of data [10], which affects the accuracy of the prediction data and introduces uncertainties to related research. Process-based models mainly investigate and simulate ecological processes occurring in plants and have extensive theoretical foundations in related fields. However, process-based models have complex structures and often simulate ideal ecological processes that deviate from actual conditions [2,11], which affects model accuracy. In addition, plant organisms involve complex and nonlinear biological and chemical mechanisms [2,12,13], which pose a great challenge for process-based models to simulate these mechanisms.

Currently, artificial intelligence (AI) algorithms have been widely applied in various fields because they can fit complex nonlinear mapping relationships between predictive and driving factors without requiring as many complicated prior assumptions as traditional models do [14]. Commonly applied machine learning models include Random Forest (RF), Support Vector Machine (SVM), and neural networks such as Long-Short Term Memory (LSTM). Tramontana et al. [15], Ichii et al. [16], Wang et al. [17], and other researchers used AI methods for tree species classification and carbon flux prediction, demonstrating the potential of AI in ecology. In recent years, Zhang et al. [18], Yuan et al. [19], Yu et al. [20], Sarkar et al. [4], and others used RF, Convolutional Neural Network (CNN), and Deep Belief Network (DBN) to predict GPP and achieved good results.

In this work, we constructed a model with spatial–temporal correlation while considering environmental factor fusion based on the GeoMAN model, a network with a multi-level attention mechanism developed by Liang et al. [21]. We trained and parameterized the algorithm with observational data on GPP from various sites and environmental driving data to extract nonlinear mapping relationships between GPP and multiple environmental factors. We designed a series of case studies to assess the performance of this deep learning model, which was based on the attention mechanism, by examining the impacts of distance, vegetation, and environmental factors on the prediction results across various flux sites. Compared with the previous applications of AI models in cross-disciplinary fields, our method not only fully utilizes the high precision of AI in prediction but also considers the prior knowledge within the relevant field to ensure that the results are both more accurate and consistent with domain knowledge.

2. Materials and Methods

2.1. Study Area

The flux sites used in this study were distributed in the Tibetan Plateau region, which is located in the alpine climate zone and has the climatic characteristics of long sunshine hours, intense sunlight, low temperatures, and scant rainfall. The regional average elevation exceeds 4000 m, the annual average temperature ranges from −5.75 to 2.57 °C, and the annual average amount of precipitation is 200–600 mm. Alpine grassland covers more than 60% of the surface area of this region [14,22], and it is a distinctive grassland ecosystem within all alpine areas in the world [23,24].

According to the *Atlas of Grassland Resources in China* (1:1,000,000) [25], alpine grasslands are subdivided into four sub-categories: alpine Kobresia meadow (KO), alpine shrub meadow (SH), alpine swamp meadow (SW), and alpine meadow steppe (AS) (Figure 1).

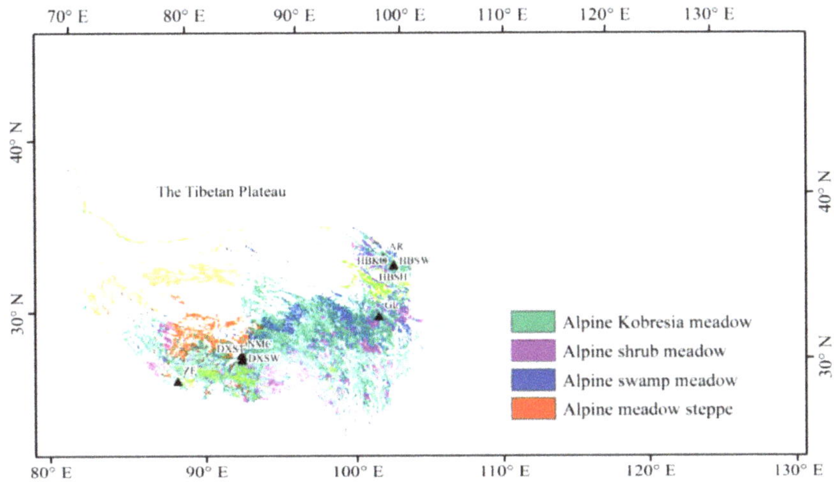

Figure 1. Geographical spread of alpine grasslands across China [14,23]. Black triangles represent the nine flux sites.

2.2. Data

2.2.1. Flux and Meteorological Data

The flux and meteorological data used in this study were collected from the China Terrestrial Ecosystem Flux Observation and Research Network (ChinaFLUX) [14,23,26], the Coordinated Observations and Integrated Research over Arid and Semi-arid China (COIRAS) [27], and the Heihe Watershed Allied Telemetry Experimental Research (HiWATER) [28], which were observed by nine flux stations distributed in the Tibetan Plateau region. The data spans from 2003 to 2014. These flux sites exemplify the broadest grassland ecosystem types, encompassing an extensive range of spatial, ecological, and weather-related circumstances [23].

Carbon flux data were processed using various methods, including triple coordinate rotation, Webb–Pearman–Leuning (WPL) correction, and outlier removal. The temporal resolution of the MODIS data was eight days, while that of temperature and photosynthetically active radiation was half an hour. The eddy covariance system was used to concurrently record these meteorological data, and any missing values were supplemented using the technique proposed by Schwalm et al. [29]. The data were then averaged and summed over eight days [14,23]. Finally, a total of 1421 site observation data points with a temporal resolution of eight days were obtained. The primary attributes of the nine flux sites in northern China's grasslands were shown in Table 1.

Table 1. Primary attributes of the nine flux sites in northern China's grasslands [14,23].

Site	Grassland Type	Latitude	Longitude	Elevation (m)	Operation Period
AR	Alpine Kobresia Meadows	38.04° N	100.46° E	3033	2014
GL		34.35° N	100.56° E	3980	2007, 2010–2011, and 2013
HBKO		37.61° N	101.31° E	3148	2003–2004
HBSH	Alpine Shrub Meadows	37.67° N	101.33° E	3293	2003–2012
DXSW	Alpine Swamp Meadows	30.47° N	91.06° E	4286	2009–2010
HBSW		37.61° N	101.33° E	3160	2004–2008 and 2010–2012
DXST	Alpine Meadow Steppes	30.5° N	91.06° E	4333	2004–2005, 2007, and 2009–2010
NMC		30.77° N	90.96° E	4730	2009
ZF		28.36° N	86.95° E	4293	2009

2.2.2. Remote Sensing Data

In this research, the remote sensing data utilized comprised the following MODIS products: normalized difference vegetation index (NDVI), enhanced vegetation index (EVI) (MOD13A2) [14,30], and surface reflectance (MOD09A1) [14,31]. The spatial resolution of the NDVI and EVI products was 1000 m and the temporal resolution was 16 days, while the spatial resolution of the surface reflectance was 500 m and the temporal resolution was 8 days. In order to acquire data with consistent spatial and temporal resolution, the quality control and data completion approaches proposed by Ma et al. [32] and Xiao et al. [33] were applied. The surface reflectance data were used to calculate the land surface water index (LSWI) [34].

2.3. Model

2.3.1. Deep Learning Model

In this study, we constructed our model based on the GeoMAN algorithm developed by Liang et al. in 2018, which was originally applied to predict air quality [21]. The GeoMAN algorithm can extract the spatial correlation of input variables and consider the influence of neighboring sites on the target site's GPP, which can help estimate the GPP of grasslands more accurately. The GeoMAN algorithm consists of an encoder and a decoder. The encoder contains a mechanism to consider the features within a site, a mechanism to consider the spatial features between sites, and an LSTM model to extract the local features of the site to be predicted and the spatial features of relevant surrounding sites. The decoder includes a temporal attention mechanism and an LSTM model, which decode the feature vector output by the encoder to predict grassland GPP.

There are complex correlations between environmental variables and GPP at each flux site. The inter-site feature attention mechanism of the GeoMAN algorithm dynamically captures the association between environmental variables and GPP within the site targeted for prediction. The inter-site feature attention mechanism for the target flux site is estimated as follows:

$$e_{k,t} = \boldsymbol{v}_0^T \tanh\left(\boldsymbol{W}_0\left[\boldsymbol{h}_{t-1}; \boldsymbol{s}_{t-1}\right] + \boldsymbol{U}_0\left[\boldsymbol{I}_{k\cdot}^0\right] + \boldsymbol{b}_0\right) \tag{1}$$

In Equation (1), $[\boldsymbol{h}_{t-1}; \boldsymbol{s}_{t-1}]$ denotes the concatenation operation in Tensorflow between $\boldsymbol{h}_{t-1}$ and $\boldsymbol{s}_{t-1}$, as they are the hidden state and the cell state of the LSTM network at time $t-1$, respectively, which contain the information of the previous $t-1$ time steps, thereby forming the long-term and short-term memory of the LSTM network. $\boldsymbol{I}_k^0$ means the k-th time series at the flux site to be predicted. $\boldsymbol{v}_0$, $\boldsymbol{W}_0$, $\boldsymbol{U}_0$, and $\boldsymbol{b}_0$ are the learnable parameters: during the learning and training process of the model, they are continuously updated according to the loss function via backpropagation. The environmental factors selected for predicting GPP in this study include temperature (Ta), photosynthetically active radiation (PAR), enhanced vegetation index (EVI), normalized difference vegetation index (NDVI), and land surface water index (LSWI). The formula for calculating the weighting values of each factor based on the Geoman model's inter-site feature attention mechanism is as follows:

$$\boldsymbol{\alpha}_{k,t} = \frac{\exp(e_{k,t})}{\sum_{j=1}^{T} \exp(e_{j,t})} \tag{2}$$

In Equation (2), $\boldsymbol{\alpha}_{k,t}$ is the weighting value of the k-th feature (one each for Ta, PAR, EVI, NDVI, and LSWI) at time t. The sum of the weighting values of all features is 1. The calculated weighting values are multiplied by the corresponding feature values to distinguish the importance of different features according to the GeoMAN model.

As shown in Figure 2, the LSTM unit is an important computational unit in the GeoMAN model. Figure 3 shows the schematic diagram of the LSTM unit. The input at time t is x_t, and c_{t-1} and h_{t-1} are the cell state and the hidden state at time $t-1$, respectively. They go through three main stages inside the LSTM unit: first, the forgetting stage, which selectively forgets the input from the previous time step; second, the selective memory stage, which selectively remembers the input from the current time step, emphasizing important parts while remembering less important parts; and third, the output stage, which outputs the new hidden state h_t and the cell state c_t and inputs them to the next time step.

Figure 2. The structure of GeoMAN adapted from Liang et al. [21].

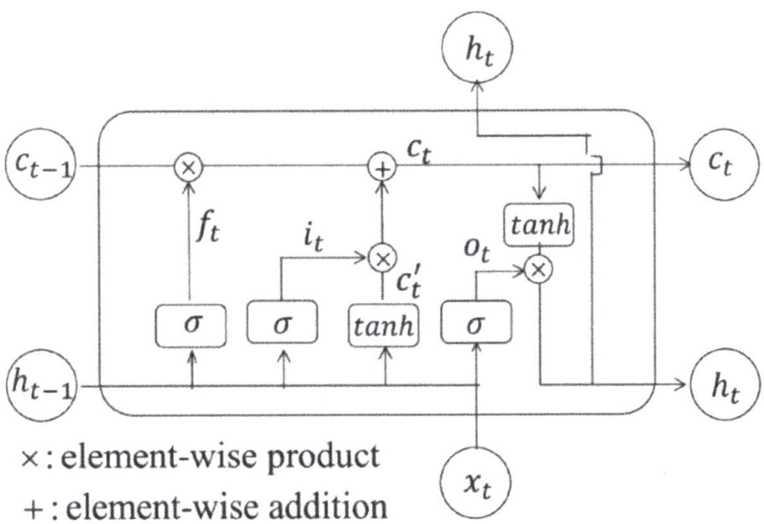

Figure 3. The structure of LSTM.

After the input data are assigned different weights by the inter-station feature attention mechanism and updated by the LSTM unit, it is essential to select pertinent time periods for GPP forecasting. It further enhances the precision of the prediction outcomes. Therefore, the GeoMAN model introduces a temporal attention mechanism. The formula for calculating the attention weight of each hidden state at a historical time step is as follows [35]:

$$u_{t,\tau}= v_d^T \tanh\left(W_d\left[h'_{\tau-1}; s'_{\tau-1}\right] + W'_d h_t + b_d\right. \tag{3}$$

$$\gamma_{t,\tau}= \frac{\exp(u_{t,\tau})}{\sum_{j=1}^{T}\exp(u_{j,\tau})} \tag{4}$$

In Equation (3), $h'_{\tau-1}$ and $s'_{\tau-1}$ are the hidden state and the cell state of the LSTM at time step $\tau - 1$. τ is the output prediction time step. v_d, W_d, U_d, and b_d are the learnable parameters. The output vector of the time attention mechanism at this time step is as follows [35]:

$$c_\tau= \sum_{t=1}^{T} \gamma_{t,\tau} h_t \tag{5}$$

2.3.2. Model Training and Evaluation

The values of each element of the flux site data used in this study have large differences. To ensure that model learning and training are not affected by this issue, each element of the data is standardized and normalized using the following formula:

$$x'= \frac{x-\mu}{\sigma} \tag{6}$$

In Equation (6), μ and σ are the mean and standard deviation of the corresponding element, respectively. The processed element data have a mean of 0 and a variance of 1, which prevents the model from being biased toward elements with large value ranges and ensures the accuracy of the model. After data pre-processing, the learning and training steps start. Since the observation data from the nine flux sites are not large in scale (a total of 1421 records), ten-fold cross-validation was applied to make full use of the data and to ensure the model's prediction performance on the whole data set. That is, for each fold, 10% of the data were taken as the test set, and the remaining 90% were used for learning and training. Then, the data used for learning and training were divided into training and validation sets at a ratio of 9:1, and the data order was randomly shuffled to avoid over-fitting. The model uses Mean Squared Error (MSE) as the loss function and the Adam optimizer to update model parameters. This process was repeated ten times to complete the predictions on all the data and evaluate the results.

This study used three common statistical indicators to evaluate model prediction accuracy: mean squared error, mean absolute error, and R-squared. The relevant formulas are as follows:

$$R^2= \left(\frac{\sum_{i=1}^{n}(y_i-\overline{y})(y'_i-\overline{y}')}{\sqrt{\sum_{i=1}^{n}(y_i-\overline{y})^2}\sqrt{\sum_{i=1}^{n}(y'_i-\overline{y}')^2}}\right)^2 \tag{7}$$

$$RMSE= \sqrt{\frac{1}{n}\sum_{i=1}^{n}(y'_i-y_i)^2} \tag{8}$$

$$MAE= \frac{1}{n}\sum_{i=1}^{n}\left|y'_i-y_i\right| \tag{9}$$

where y'_i and y_i are the predicted and observed values of GPP, respectively; $\overline{y}'$ and $\overline{y}$ are the mean values of y'_i and y_i, respectively; and n is the number of observation samples.

3. Case Analysis

In this section, we introduced the basic idea of our experimental design. We tested the performance of different models on the aggregate data to prove that deep learning models have highly accurate prediction capabilities. Moreover, considering that the model predictions must conform to basic ecological knowledge, we were required to conduct multiple experiments by controlling the spatial distance of flux sites, vegetation types, and environmental factors.

3.1. Prediction Accuracy with All Factors

3.1.1. Comparison of Model Performance with the Use of All Data

In this analysis, we used the data from all flux sites to train the Random Forest (RF), Support Vector Machine (SVM), Deep Belief Network (DBN), and GeoMAN models. According to the ten-fold cross-validation results of all the models, the relationship between predicted GPP and observed GPP is shown in Figure 4 and Table 2. It is obvious that there are different training effects between the four models. The Random Forest model has the lowest prediction accuracy, with a prediction RMSE of 0.954 g Cm^{-2} d^{-1}, MAE of 0.553 g Cm^{-2} d^{-1}, and R^2 of 0.810; the Deep Belief Network model achieves a certain improvement in prediction accuracy compared to the Random Forest model, with a prediction RMSE of 0.912 g Cm^{-2} d^{-1}, MAE of 0.559 g Cm^{-2} d^{-1}, and R^2 of 0.827; the Support Vector Machine model has similar prediction accuracy to the Deep Belief Network model, with a prediction RMSE of 0.910 g Cm^{-2} d^{-1}, MAE of 0.571 g Cm^{-2} d^{-1}, and R^2 of 0.827; and the GeoMAN model has the highest prediction accuracy with a prediction RMSE of 0.788 g Cm^{-2} d^{-1}, MAE of 0.440 g Cm^{-2} d^{-1}, and R^2 of 0.870, which indicates that the GeoMAN model could fit the GPP values of the nine flux sites in the Tibetan Plateau better than the other three models.

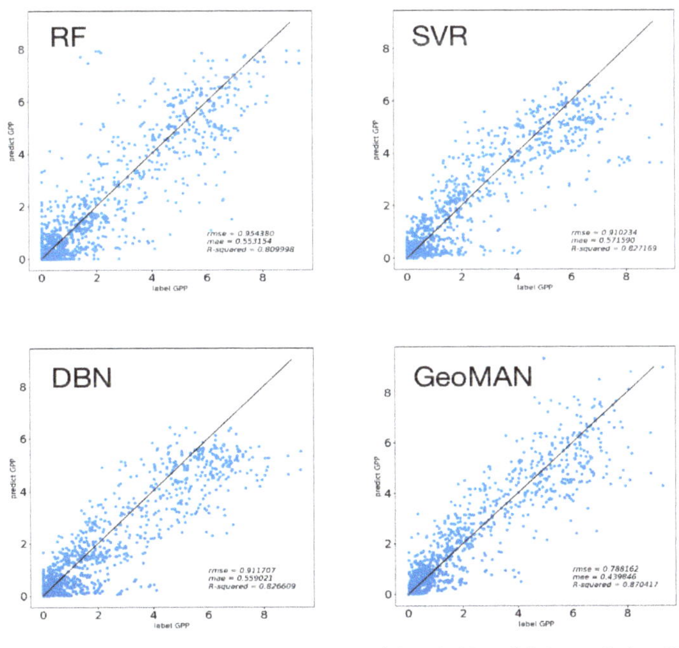

Figure 4. Performance of RF, SVR, DBN, and GeoMAN models in predicting GPP.

Table 2. Results of the four models with the use of all data.

Model	RF	SVR	DBN	GeoMAN
RMSE	0.954	0.910	0.912	0.788
MAE	0.553	0.571	0.559	0.440
R^2	0.810	0.827	0.827	0.870

3.1.2. Performance of Single Flux Site

As shown in Figure 1, the distribution map of the nine flux sites indicates that different sites have different vegetation and climate conditions. Therefore, it is necessary to test the GPP prediction accuracy of the GeoMAN model at different flux sites.

1. Test site GPP against remaining sites

We took the target flux site as the test data and the remaining sites as the training data, for a total of nine sites being tested. As shown in Figure 5, there is a large difference in performance among the different flux sites. The possible reasons for this difference are (1) some sites have different vegetation types from the target site, and (2) some sites have different climate conditions from the target site due to their long distance.

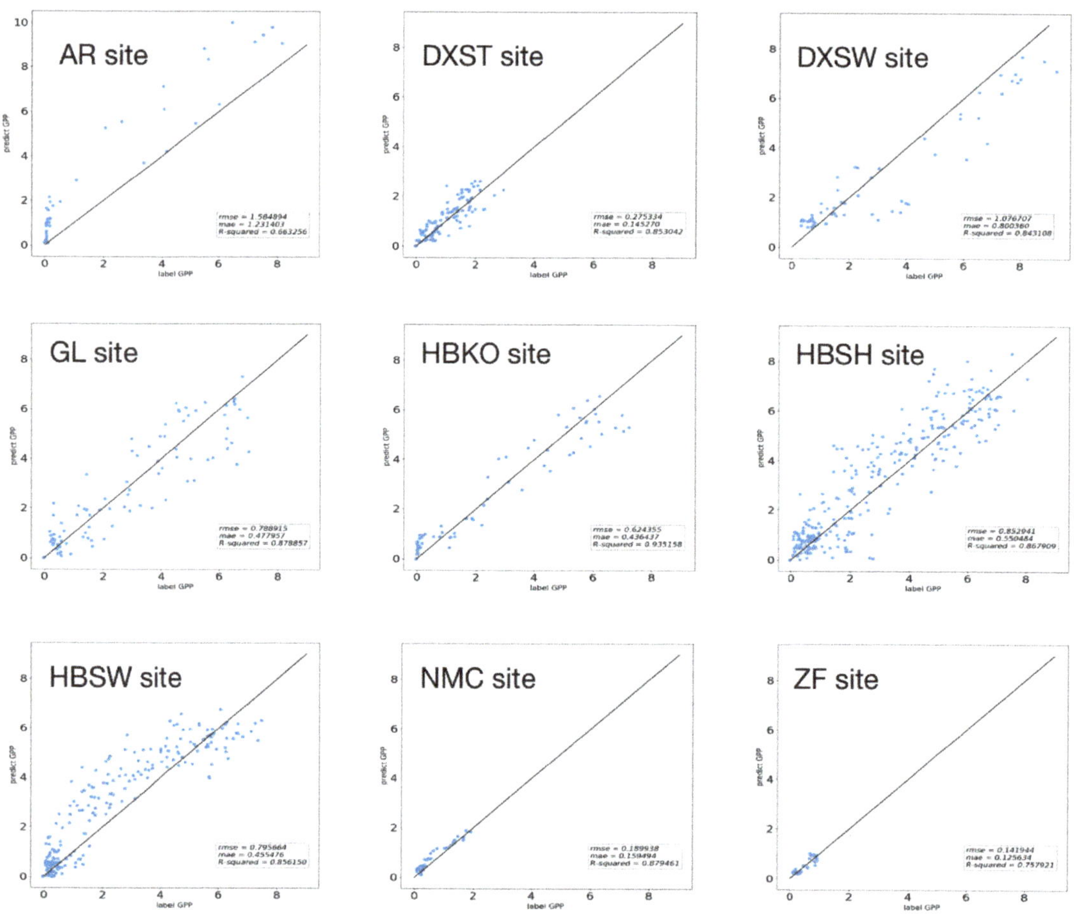

Figure 5. Predicted GPP vs. labeled GPP at a single site.

2. Test site GPP against the other sites at distance of 500 km and 100 km

The distances between each flux site were calculated using the Haversine formula based on their latitude and longitude. The calculation results are shown in Table 3.

Table 3. Distances between the flux sites (km).

	AR	DXST	DXSW	GL	HBKO	HBSH	HBSW	NMC	ZF
AR	/	1202.5	1204.9	410.4	88.7	86.8	90.1	1187.5	1651.6
DXST	1202.5	/	3.3	988.5	1230.1	1235.6	1231.6	31.5	463.7
DXSW	1204.9	3.3	/	990.1	1232.4	1237.8	1233.8	34.7	462.0
GL	410.4	988.5	990.1	/	368.7	375.6	369.1	983.3	1452.1
HBKO	88.7	1230.1	1232.4	368.7	/	6.9	1.8	1217.1	1685.3
HBSH	86.8	1235.6	1237.8	375.6	6.9	/	6.7	1222.44	1690.5
HBSW	90.1	1231.6	1233.8	369.1	1.8	6.7	/	1218.6	1686.8
NMC	1187.5	31.5	34.7	983.3	1217.1	1222.4	1218.6	/	471.4
ZF	1651.6	463.7	462.0	1452.1	1685.3	1690.5	1686.8	471.4	/

Based on the results in Table 3, each target flux site was predicted using the flux sites within 500 km and 100 km as the training data. There are no flux sites within 100 km of the GL and ZF sites, so they were not included in the prediction results for sites within 100 km. The prediction results are shown in Figures 6 and 7. According to the results, the accuracy of the AR site increases as the range of selected sites decreases, while the DXST, NMC, and ZF sites show a decreasing trend in accuracy as the range of selected sites decreases. The overall accuracy of the DXSW site is lower than when using all sites for prediction, but it shows a rebound trend as the range decreases. According to Table 1, the AR site with an increasing trend has alpine Kobresia meadow as its vegetation type, while the DXST, NMC, and ZF sites with a decreasing trend have alpine meadow steppe as their vegetation type. It can be speculated that the prediction effect of each site is related to the vegetation type of the other selected sites.

3. Selecting training data according to vegetation type

As shown in Table 1, the training data for each site to be predicted comes from the other flux sites with the same vegetation type. The final predictions are shown in Figure 8. It is obvious that selecting the training data based on vegetation type has higher overall prediction accuracy compared to selecting the training data based on distance (no corresponding prediction results are available for the HBSH site because it has a different vegetation type compared to the other flux sites). Next, we combined the results of the previous three experiments to examine the effect of selecting training data under different conditions on prediction accuracy, and the combined results are shown in Table 4.

As shown in Table 4, using vegetation type as the training data in the screening mechanism is better than using site distance as the training data from an overall perspective. However, from a single-site perspective, the GL and HBKO sites show a decreasing trend in accuracy. This is because for the AR and HBKO sites, which have significantly less data than other sites and share the same vegetation type as the GL site, the training data are insufficient, leading to a decrease in the prediction accuracy at the GL site. The HBKO site, which has always maintained an R-squared value above 0.9 from an overall perspective, does not have much room for accuracy improvement. At the same time, we compared the prediction results of the training data without screening and with screening for vegetation type. Although some sites have lower accuracy, the AR site shows a significant improvement in accuracy; as a result, the overall prediction accuracy of the latter method is not lower than that of the former method. This result shows that increasing the amount of training data with the same vegetation type can achieve equally good results as increasing the overall amount of training data.

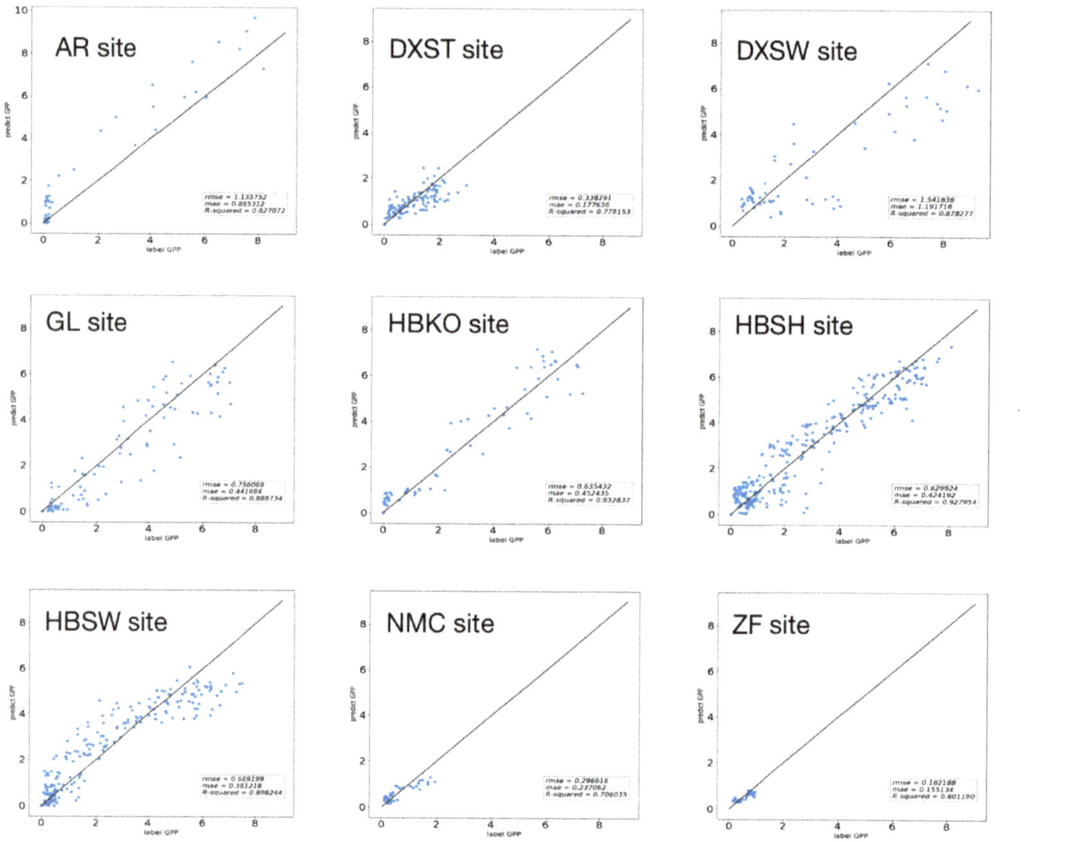

Figure 6. Predicted GPP vs. the labeled GPP at a single site (500 km).

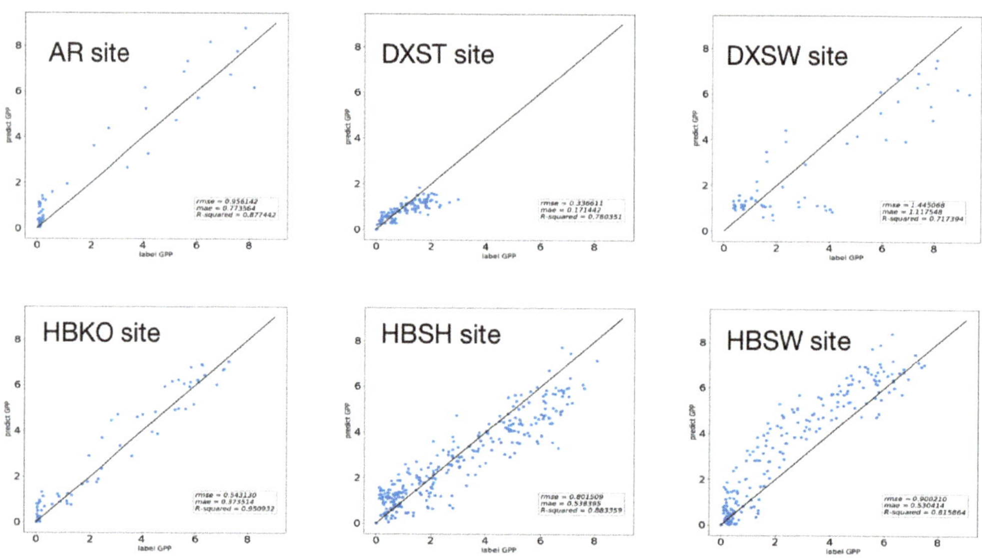

Figure 7. *Cont.*

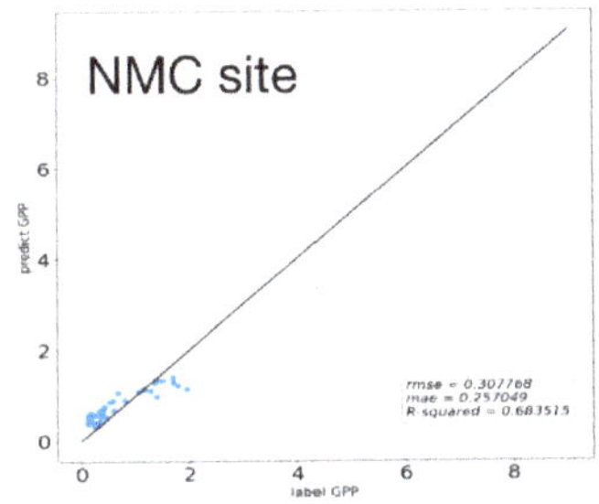

Figure 7. Predicted GPP vs. labeled GPP at a single site (100 km).

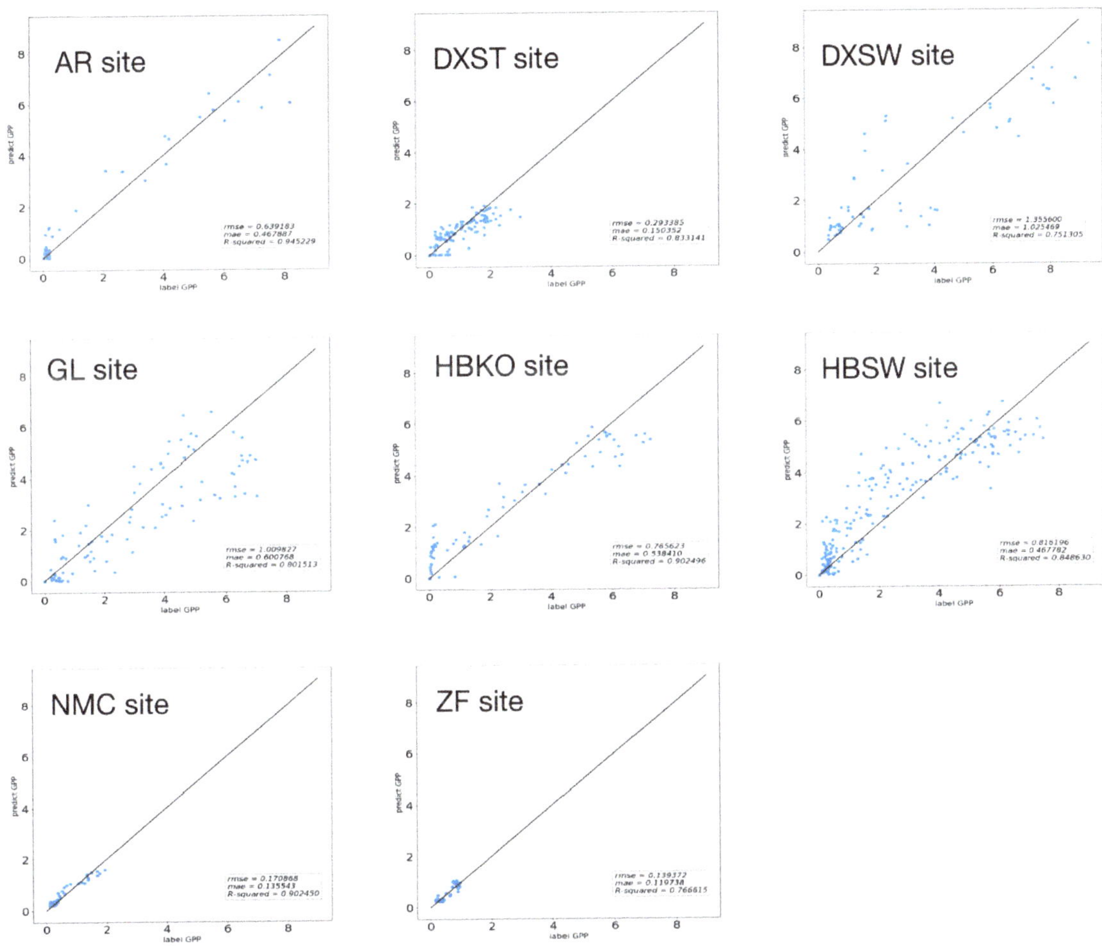

Figure 8. Predicted GPP vs. labeled GPP at a single site based on vegetation type.

Table 4. Results comparing all previous experiments.

Site	AR	DXST	DXSW	GL	HBKO	HBSH	HBSW	NMC	ZF
R^2	0.663	0.853	0.843	0.879	0.935	0.868	0.856	0.879	0.758
R^2 (500 km)	0.827	0.778	0.678	0.889	0.933	0.928	0.898	0.706	0.601
R^2 (100 km)	0.877	0.780	0.717	/	0.951	0.883	0.816	0.683	/
R^2 (vt)	0.945	0.833	0.751	0.801	0.902	/	0.849	0.902	0.767

Numbers in red indicate that the prediction results for the corresponding site have increased in accuracy compared to the previous results without setting a distance range; numbers in blue indicate that the results have decreased in accuracy; and numbers in black indicate that the results have no significant changes.

3.2. *Prediction Accuracy with Factor Ablation*

In Section 3.1, the effects of vegetation type and distance between the flux sites on GPP prediction accuracy were then investigated. The training data included temperature (Ta), photosynthetically active radiation (PAR), enhanced vegetation index (EVI), normalized difference vegetation index (NDVI), and land surface water index (LSWI). In this analysis, a feature ablation experiment was conducted to explore the influence of each factor on GPP prediction accuracy.

3.2.1. Test Site GPP without Ta

In this experiment, all Ta data were deleted from the training data, which were then trained for each site flux. The final prediction results are shown in Figure 9. Compared to the prediction results without any feature ablation of the training data, the prediction accuracy of the AR site is significantly improved, while that of the DXST, DXSW, NMC, and ZF sites is greatly reduced. The prediction accuracy of the GL, HBKO, HBSH, and HBSW sites does not change noticeably.

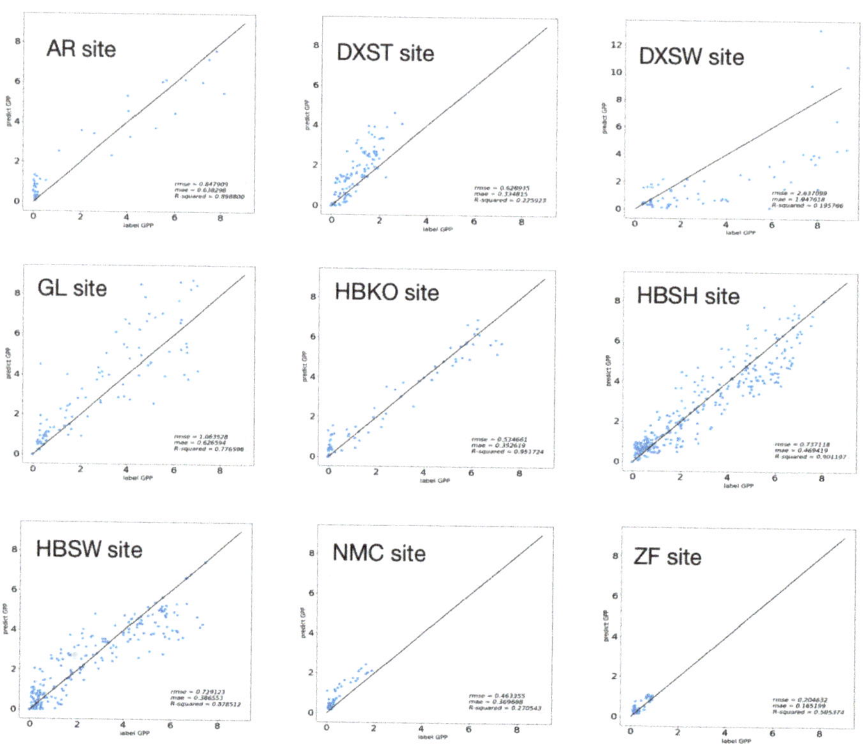

Figure 9. Predicted GPP vs. labeled GPP at a single site (no Ta).

3.2.2. Test Site GPP without PAR

In this experiment, all Par data were deleted from the training data, which were then trained for each site flux. The final prediction results are shown in Figure 10. Compared to the prediction results without any feature ablation of the training data, the prediction accuracy trends of the sites are similar to those obtained after removing Ta. The prediction accuracy of the AR site has significantly improved, but not as much as after the removal of Ta. The prediction accuracy of the DXST, DXSW, NMC, and ZF sites is sharply reduced. For the DXST site, the decrease is greater than that after removing Ta, whereas the DXSW, NMC, and ZF sites show some recovery but still perform worse than without any feature ablation. The prediction accuracy of the GL, HBKO, HBSH, and HBSW sites as a whole is lower than that after removing Ta, though it does not change appreciably.

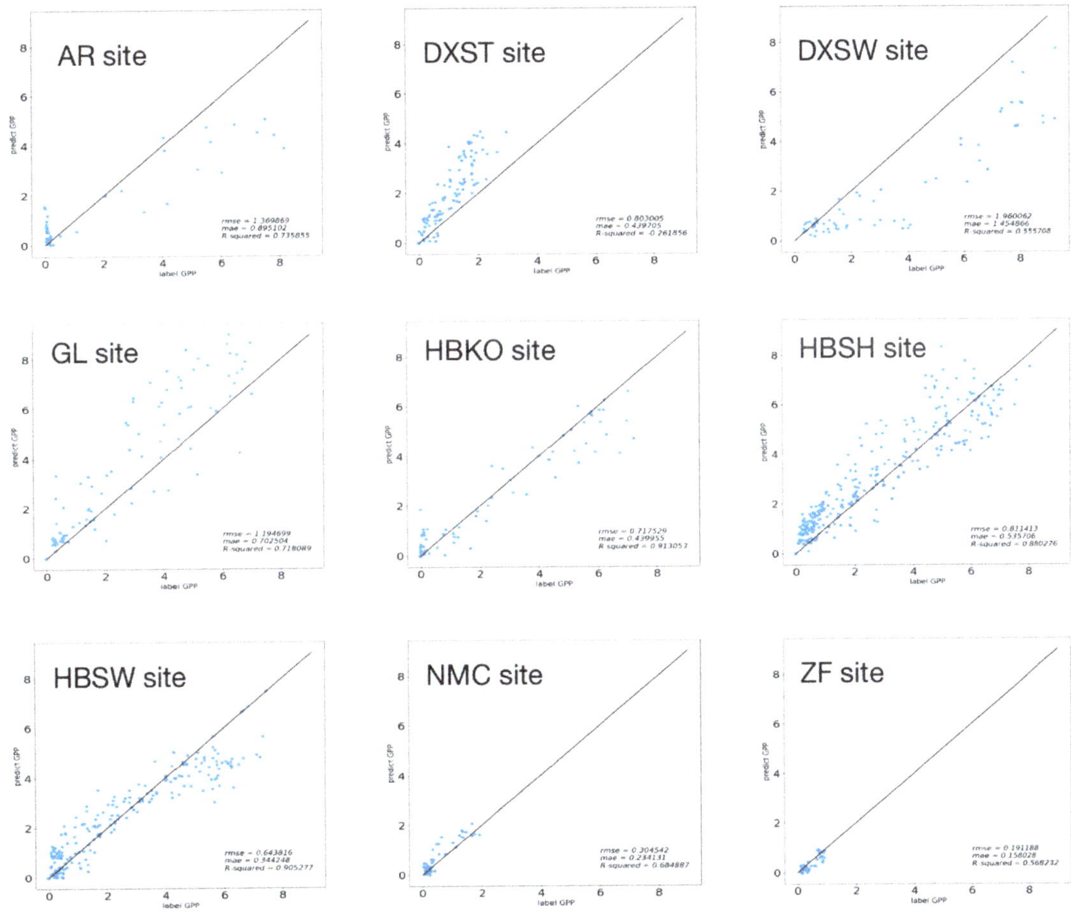

Figure 10. Predicted GPP vs. labeled GPP at a single site (no PAR).

3.2.3. Test Site GPP without EVI

In this experiment, all EVI data were deleted from the training data, which were then trained for each site flux. The final prediction results are shown in Figure 11. Compared to the prediction results without any feature ablation of the training data, the prediction accuracy of the AR site is still significantly improved and higher than that after removing Par but lower than that after removing Ta. The prediction accuracy of the DXST, DXSW, NMC, and ZF sites is similar to that after the removal of Ta and Par, with a sharp decrease.

The decrease is larger for the NMC and ZF sites than for the others. The prediction accuracy of the GL site shows a continuous decline compared to that after the removal of Ta and Par. The prediction accuracy of the HBKO, HBSH, and HBSW sites does not change noticeably.

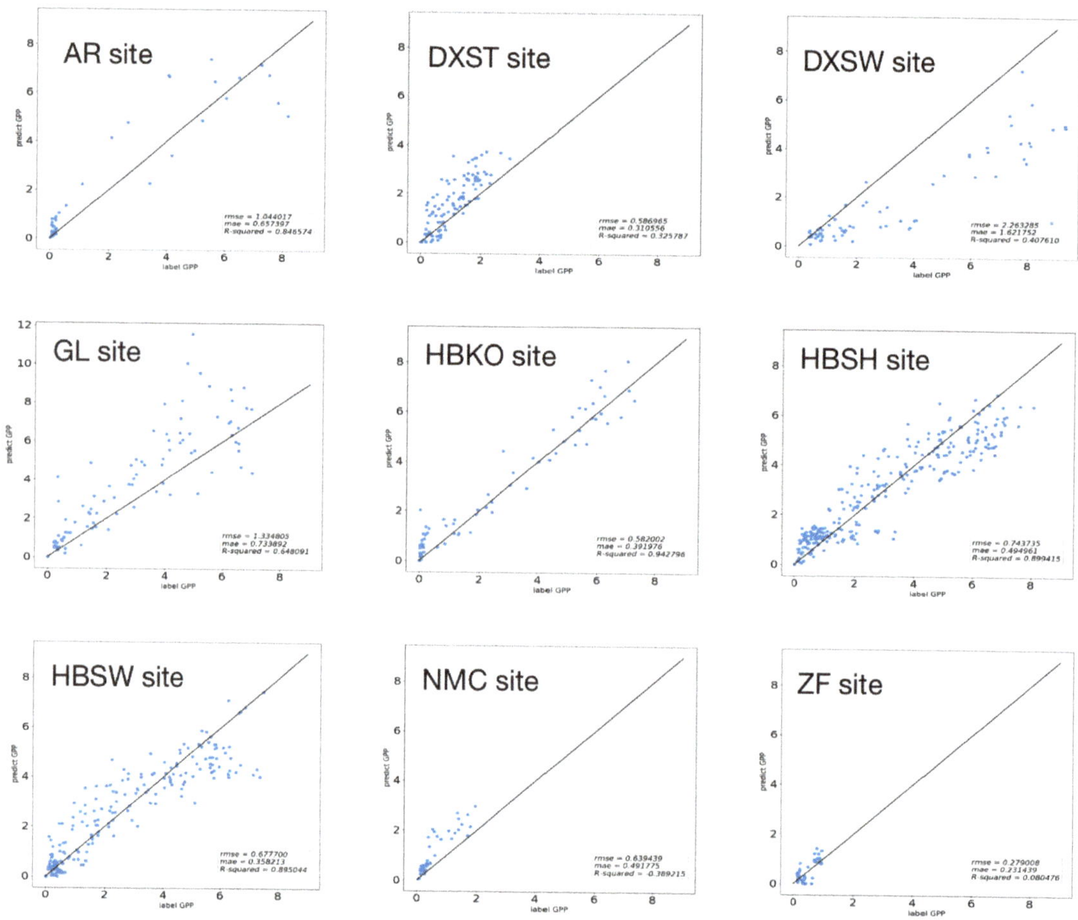

Figure 11. Predicted GPP vs. labeled GPP at a single site (no EVI).

3.2.4. Test Site GPP without NDVI

In this experiment, all NDVI data were deleted from the training data, which were then trained for each site flux. The final prediction results are shown in Figure 12. Compared to the prediction results without any feature ablation of the training data, the prediction accuracy of the AR site is significantly improved and higher than that after removing Ta, Par, and EVI. The prediction accuracy of the DXST, DXSW, and ZF sites decreases, which is consistent with the results of the previous ablation experiments. For the ZF site, the prediction accuracy is only better than that after removing Ta and EVI. The NMC site has an abnormal increase in accuracy. The prediction accuracy of the GL site is higher than that after removing Ta, Par, and EVI. The prediction accuracy of the HBKO, HBSH, and HBSW sites does not change noticeably.

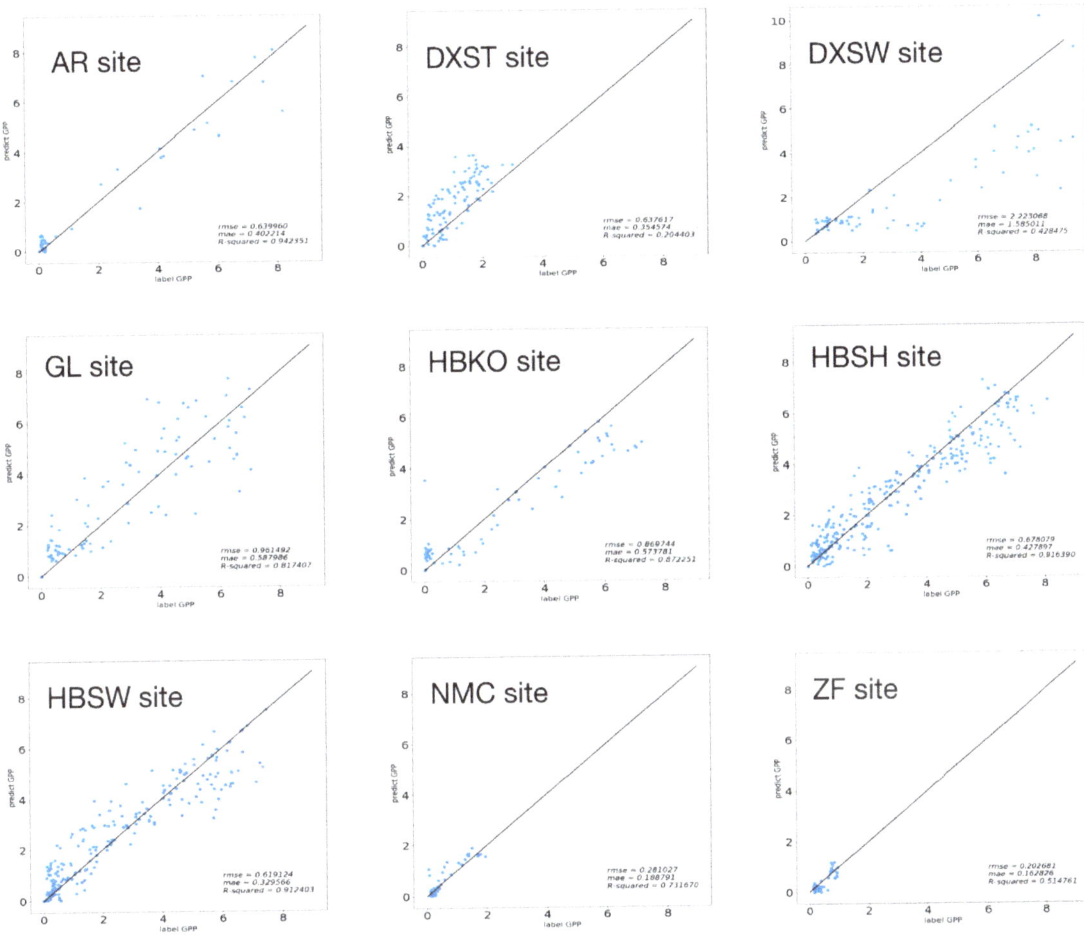

Figure 12. Predicted GPP vs. labeled GPP at a single site (no NDVI).

3.2.5. Test Site GPP without LSWI

In this experiment, all LSWI data were deleted from the training data, which were then trained for each site flux. The final prediction results are shown in Figure 13. Compared to the prediction results without any feature ablation of the training data, the prediction accuracy of the AR site improves significantly and is only lower than that after removing NDVI. The DXST, DXSW, NMC, and ZF sites have similar prediction accuracy as in the previous ablation experiments, with a large decrease in accuracy, and the prediction accuracy of the ZF site is only higher than that after removing EVI. The GL site shows a slight decrease in accuracy, while the HBKO, HBSH, and HBSW sites show no obvious changes in prediction accuracy.

3.2.6. Summary of Factor Ablation Experiments

To observe the influence of different features on the prediction accuracy of each site more intuitively, we summarized all the feature ablation experiment results, as shown in Table 5.

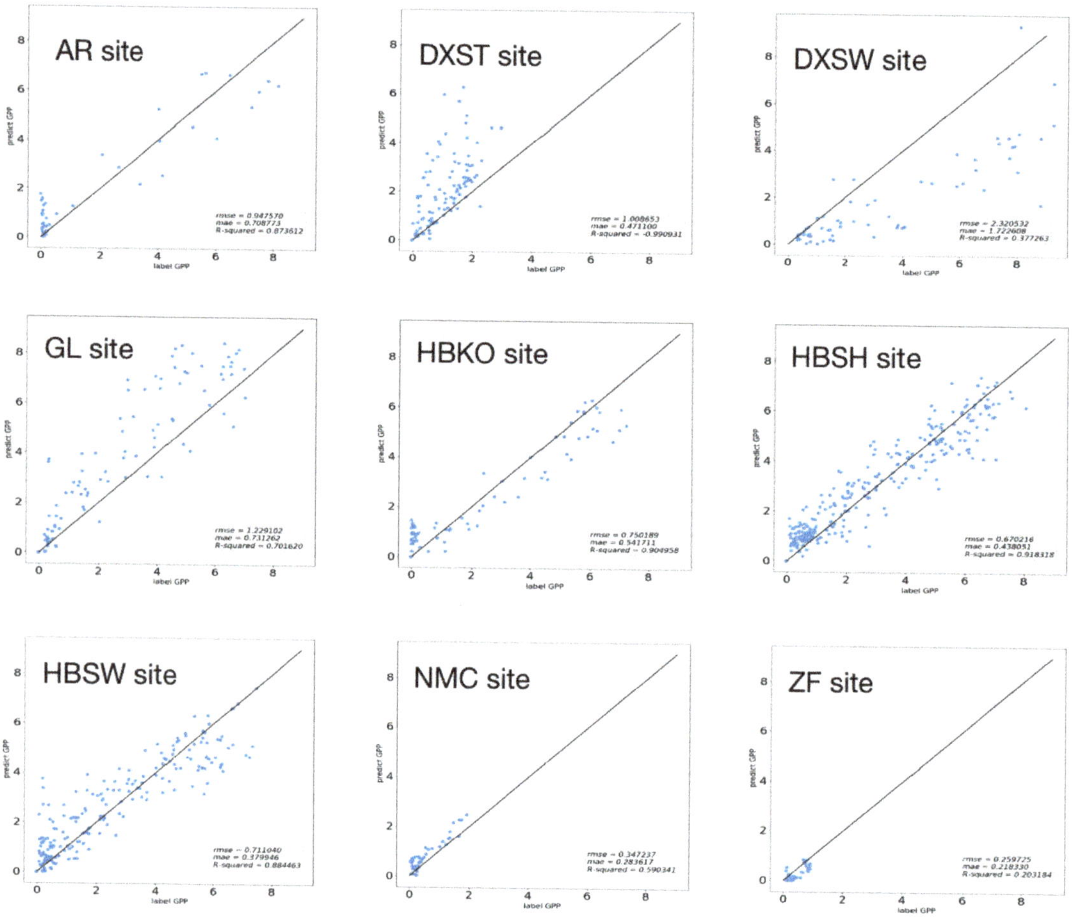

Figure 13. Predicted GPP vs. labeled GPP at a single site (no LSWI).

Table 5. Results of all factor ablation experiments.

Site	AR	DXST	DXSW	GL	HBKO	HBSH	HBSW	NMC	ZF
R^2	0.663	0.853	0.843	0.879	0.935	0.868	0.856	0.879	0.758
R^2 (no-Ta)	0.899	0.226	0.196	0.777	0.952	0.901	0.879	0.271	0.505
R^2 (no-PAR)	0.736	−0.262	0.556	0.718	0.913	0.880	0.905	0.685	0.568
R^2 (no-EVI)	0.847	0.326	0.408	0.648	0.943	0.899	0.895	−0.385	0.080
R^2 (no-NDVI)	0.942	0.204	0.428	0.817	0.872	0.916	0.912	0.732	0.515
R^2 (no-LSWI)	0.874	−0.991	0.377	0.702	0.905	0.918	0.884	0.590	0.203

Numbers in red indicate that the prediction results for the corresponding site have increased in accuracy compared to the previous results without setting a distance range; numbers in blue indicate that the results have decreased in accuracy; and numbers in black indicate that the results have no significant changes.

Table 5 indicates that removing any feature from the AR site would result in a significant improvement in accuracy, with the largest improvement obtained after removing NDVI. The removal of any feature for the DXST, DXSW, and ZF sites would lead to a degradation of accuracy, with the DXST site showing a large accuracy decline and the lowest accuracy after removing LSWI. The DXSW site also shows a decline in accuracy, although it is smaller than that of the DXST site. The ZF site has a noticeable decline in accuracy after removing EVI. The NMC site has an abnormal increase in accuracy after removing NDVI and a decline after removing other features except NDVI. The GL site is insensitive to the removal of Ta or NDVI and shows slight decreases in accuracy after removing other features in addition to Ta and NDVI. The HBKO, HBSH, and HBSW sites are insensitive to the removal of any feature and have no obvious changes in accuracy.

4. Conclusions

In this work, we used satellite remote sensing data and flux site observation data to introduce the GeoMAN model based on an encoder–decoder framework with an attention mechanism for site features, and we obtained good results. According to the experiments on training data selection based on distance and vegetation type, we found that both distance and vegetation type had an impact on GPP prediction results, with vegetation type having a larger impact. Through the feature ablation experiments, we found that different sites showed sensitivity to different factors, with the site located in the alpine swamp meadow being insensitive to changes in environmental factors, while the site located in the alpine meadow steppe showed a different trend since the GPP prediction accuracy decreased sharply with the changes in environmental factors. The GPP prediction accuracy of the site located in the alpine Kobresia meadow also varied with environmental factor changes but was more stable than the other sites. The results of this work show that deep learning models have high accuracy when simulating site-scale GPP and, to some extent, reflect the correlation between a target site's GPP and other sites' distances, vegetation types, and meteorological factors. Our work could be used in the prediction of other factors, for example, AGB (Above Ground Biomass) and RE (Ecosystem Respiration). However, this work has some limitations. Firstly, we do not consider some factors that have an influence on productivity in Tibetan grasslands, such as soil development and drought regimes. Secondly, the data we used in this work only cover a partial area of the Tibetan Plateau region, and this introduces constraints to regional GPP assessment. In our future work, we will add more factors to the training data, for example, soil pH, soil fertility, and soil organic matter (SOM), since high soil pH and a lack of soil fertility limit plant productivity [36], and SOM is able to enhance alpine grassland productivity by improving the soil structure, aggregates, and cation-exchange capacity (CEC) under high aridity conditions [37]. Moreover, ecological factors, such as growing and non-growing seasons, will be considered, and larger regional-scale data will be used in future training and learning processes. These improvements will help us perform more accurate and larger-scale GPP simulations.

Author Contributions: Conceptualization, Q.Y. and Y.W.; methodology, Q.Y. and N.N.; software, Q.Y.; validation, Q.Y., Z.W. and M.W.; formal analysis, R.C.; investigation, Q.Y.; resources, X.R.; data curation, Q.Y. and X.W.; writing—original draft preparation, Q.Y.; writing—review and editing, N.N. and X.W.; visualization, Q.Y. and W.L.; supervision, Y.W.; project administration, Q.Y.; funding acquisition, N.N. and Y.W. All authors have read and agreed to the published version of the manuscript.

Funding: This research was funded by the National Key Research and Development Program of China (Grant No. 2021YFF0703902).

Institutional Review Board Statement: Not applicable.

Informed Consent Statement: Not applicable.

Data Availability Statement: Not applicable.

Conflicts of Interest: The authors declare no conflict of interest.

References

1. Williams, M.; Rastetter, E.B.; Fernandes, D.N.; Goulden, M.L.; Shaver, G.R.; Johnson, L.C. Predicting gross primary productivity in terrestrial ecosystems. *Ecol. Appl.* **1997**, *7*, 882–894. [CrossRef]
2. Zhang, X.; Wang, H.; Yan, H.; Ai, J. Analysis of spatio-temporal changes of gross primary productivity in China from 2001 to 2018 based on Romote Sensing. *Acta Ecol. Sin.* **2021**, *41*, 6351–6362.
3. Piao, S.L.; Sitch, S.; Ciais, P.; Friedlingstein, P.; Peylin, P.; Wang, X.H.; Ahlström, A.; Anav, A.; Canadell, J.G.; Cong, N.; et al. Evaluation of terrestrial carbon cycle models for their response to climate variability and to CO_2 trends. *Glob. Chang. Biol.* **2013**, *19*, 2117–2132. [CrossRef]
4. Sarkar, D.P.; Shankar, B.U.; Parida, B.R. Machine Learning Approach to Predict Terrestrial Gross Primary Productivity using Topographical and Remote Sensing Data. *Ecol. Inform.* **2022**, *70*, 101697. [CrossRef]
5. Lee, B.; Kim, N.; Kim, E.-S.; Jang, K.; Kang, M.; Lim, J.-H.; Cho, J.; Lee, Y. An Artificial Intelligence Approach to Predict Gross Primary Productivity in the Forests of South Korea Using Satellite Remote Sensing Data. *Forests* **2020**, *11*, 1000. [CrossRef]
6. Kang, M.; Kim, J.; Kim, H.S.; Thakuri, B.M.; Chun, J.H. On the nighttime correction of CO_2 flux measured by eddy covariance over temperate forests in complex terrain. *Korean J. Agric. For. Meteorol.* **2014**, *16*, 233–245. (In Korean with English abstract) [CrossRef]
7. Running, S.W.; Nemani, R.R.; Heinsch, F.A.; Zhao, M.S.; Reeves, M.; Hashimoto, H. A continuous satellite-derived measure of global terrestrial primary production. *Bioscience* **2004**, *54*, 547–560. [CrossRef]
8. Xiao, X.M.; Zhang, Q.Y.; Braswell, B.; Urbanski, S.; Boles, S.; Wofsy, S.; Moore, B., III; Ojima, D. Modeling gross primary production of temperate deciduous broadleaf forest using satellite images and climate data. *Remote Sens. Environ.* **2004**, *91*, 256–270. [CrossRef]
9. Yuan, W.P.; Liu, S.G.; Zhou, G.S.; Zhou, G.Y.; Tieszen, L.L.; Baldocchi, D.; Bernhofer, C.; Gholz, H.; Goldstein, A.H.; Goulden, M.L.; et al. Deriving a light use efficiency model from eddy covariance flux data for predicting daily gross primary production across biomes. *Agric. For. Meteorol.* **2007**, *143*, 189–207. [CrossRef]
10. Reeves, M.C.; Zhao, M.; Running, S.W. Usefulness and limits of MODIS GPP for estimating wheat yield. *Int. J. Remote Sens.* **2007**, *26*, 1403–1421. [CrossRef]
11. Beer, C.; Reichstein, M.; Tomelleri, E.; Ciais, P.; Jung, M.; Carvalhais, N.; Rödenbeck, C.; Arain, M.A.; Baldocchi, D.; Bonan, G.B.; et al. Terrestrial gross carbon dioxide uptake: Global distribution and covariation with climate. *Science* **2010**, *329*, 834–838. [CrossRef]
12. Schindler, D.E.; Hilborn, R. Prediction, precaution, and policy under global change. *Science* **2015**, *347*, 953–954. [CrossRef]
13. Ye, H.; Beamish, R.J.; Glaser, S.M.; Grant, S.C.H.; Hsieh, C.H.; Richards, L.J.; Schnute, J.T.; Sugihara, G. Equation-free mechanistic ecosystem forecasting using empirical dynamic modeling. *Proc. Natl. Acad. Sci. USA* **2015**, *112*, E1569–E1576. [CrossRef]
14. Zhu, X.; He, H.; Ma, M.; Ren, X.; Zhang, L.; Zhang, F.; Li, Y.; Shi, P.; Chen, S.; Wang, Y.; et al. Estimating Ecosystem Respiration in the Grasslands of Northern China Using Machine Learning: Model Evaluation and Comparison. *Sustainability* **2020**, *12*, 2099. [CrossRef]
15. Tramontana, G.; Ichii, K.; Camps-Valls, G.; Tomelleri, E.; Papale, D. Uncertainty analysis of gross primary production upscaling using random forests, remote sensing and eddy covariance data. *Remote Sens. Environ.* **2015**, *168*, 360–373. [CrossRef]
16. Ichii, K.; Ueyama, M.; Kondo, M.; Saigusa, N.; Kim, J.; Alberto, M.C.; Ardö, J.; Euskirchen, E.S.; Kang, M.; Hirano, T.; et al. New data-driven estimation of terrestrial CO_2 fluxes in Asia using a standardized database of eddy covariance measurements, remote sensing data, and support vector regression. *J. Geophys. Res. Biogeosci.* **2017**, *122*, 767–795. [CrossRef]
17. Wang, X.; Yao, Y.; Zhao, S.; Jia, K.; Zhang, X.; Zhang, Y.; Zhang, L.; Xu, J.; Chen, X. MODIS-based estimation of terrestrial latent heat flux over North America using three machine learning algorithms. *Remote Sens.* **2017**, *9*, 1326. [CrossRef]
18. Zhang, K.; Liu, N.; Gao, S.; Zhao, S. Data-Driven Estimation of Gross Primary Production. *Remote Sens. Technol. Appl.* **2020**, *35*, 943–949. [CrossRef]
19. Yuan, D.; Zhang, S.; Li, H.; Zhang, J.; Yang, S.; Bai, Y. Improving the Gross Primary Productivity Estimate by Simulating the Maximum Carboxylation Rate of the Crop Using Machine Learning Algorithms. *IEEE Trans. Geosci. Remote Sens.* **2022**, *60*, 4413115. [CrossRef]
20. Yu, T.; Zhang, Q.; Sun, R. Comparison of Machine Learning Methods to Up-Scale Gross Primary Production. *Remote Sens.* **2021**, *13*, 2448. [CrossRef]
21. Liang, Y.; Ke, S.; Zhang, J.; Yi, X.; Zheng, Y. Geoman: Multi-level attention networks for geo-sensory time series prediction. In Proceedings of the Twenty-Seventh International Joint Conference on Artificial Intelligence, Stockholm, Sweden, 13–19 July 2018.
22. Zhang, J.-W. *Vegetation of Xizang (Tibet)*; Science Press: Beijing, China, 1988.
23. Liu, W.; He, H.; Wu, X.; Ren, X.; Zhang, L.; Zhu, X.; Feng, L.; Lv, Y.; Chang, Q.; Xu, Q.; et al. Spatiotemporal Changes and Driver Analysis of Ecosystem Respiration in the Tibetan and Inner Mongolian Grasslands. *Remote Sens.* **2022**, *14*, 3563. [CrossRef]
24. Ge, R.; He, H.; Ren, X.; Zhang, L.; Li, P.; Zeng, N.; Yu, G.; Zhang, L.; Yu, S.-Y.; Zhang, F.; et al. A Satellite-Based Model for Simulating Ecosystem Respiration in the Tibetan and Inner Mongolian Grasslands. *Remote Sens.* **2018**, *10*, 149. [CrossRef]
25. Su, D. *The Atlas of Grassland Resources of China (1:1,000,000)*; Press of Map: Beijing, China, 1993. (In Chinese)
26. Yu, G.-R.; Wen, X.-F.; Sun, X.-M.; Tanner, B.D.; Lee, X.; Chen, J.-Y. Overview of ChinaFLUX and evaluation of its eddy covariance measurement. *Agric. For. Meteorol.* **2006**, *137*, 125–137. [CrossRef]

27. Wang, H.; Jia, G.; Fu, C.; Feng, J.; Zhao, T.; Ma, Z. Deriving maximal light use efficiency from coordinated flux measurements and satellite data for regional gross primary production modeling. *Remote Sens. Environ.* **2010**, *114*, 2248–2258. [CrossRef]
28. Li, X.; Cheng, G.D.; Liu, S.M.; Xiao, Q.; Ma, M.G.; Jin, R.; Che, T.; Liu, Q.H.; Wang, W.Z.; Qi, Y.; et al. Heihe watershed allied telemetry experimental research (hiwater): Scientific objectives and experimental design. *Bull. Am. Meteorol. Soc.* **2013**, *94*, 1145–1160. [CrossRef]
29. Schwalm, C.R.; Williams, C.A.; Schaefer, K.; Anderson, R.; Arain, M.A.; Baker, I.; Barr, A.; Black, T.A.; Chen, G.; Chen, J.M.; et al. A model-data intercomparison of CO_2 exchange across North America: Results from the North American Carbon Program site synthesis. *J. Geophys. Res. Biogeosci.* **2010**, *115*, G3. [CrossRef]
30. Huete, A.; Didan, K.; Miura, T.; Rodriguez, E.P.; Gao, X.; Ferreira, L.G. Overview of the radiometric and biophysical performance of the MODIS vegetation indices. *Remote Sens. Environ.* **2002**, *83*, 195–213. [CrossRef]
31. Vermote, E.; Vermeulen, A. *MODIS Algorithm Technical Background Document, Atmospheric Correction Algorithm: Spectral Reflectances (MOD09)*; NASA Contract NAS5-96062; University of Maryland: College Park, MD, USA, 1999.
32. Ma, M.G.; Veroustraete, F. Reconstructing pathfinder AVHRR land NDVI time-series data for the Northwest of China. *Adv. Space Res. Ser.* **2006**, *37*, 835–840. [CrossRef]
33. Xiao, J.F.; Zhuang, Q.L.; Baldocchi, D.D.; Law, B.E.; Richardson, A.D.; Chen, J.Q.; Oren, R.; Starr, G.; Noormets, A.; Ma, S.Y.; et al. Estimation of net ecosystem carbon exchange for the conterminous United States by combining MODIS and AmeriFlux data. *Agric. For. Meteorol.* **2008**, *148*, 1827–1847. [CrossRef]
34. Xiao, X.; Hollinger, D.; Aber, J.; Goltz, M.; Davidson, E.A.; Zhang, Q.; Moore, B. Satellite-based modeling of gross primary production in an evergreen needleleaf forest. *Remote Sens. Environ.* **2004**, *89*, 519–534. [CrossRef]
35. Shi, M.; Wang, J.; Yin, R.; Zhang, P. Short-Term Photovoltaic Power Forecast Basedon Grey Relational Analysis and GeoMAN Model. *Trans. China Electrotech. Soc.* **2021**, *36*, 2298–2305.
36. Zhao, Y.; Wang, X.; Jiang, S.; Xiao, J.; Li, J.; Zhou, X.; Liu, H.; Hao, Z.; Wang, K. Soil development mediates precipitation control on plant productivity and diversity in alpine grasslands. *Geoderma* **2022**, *412*, 115721. [CrossRef]
37. Zhao, Y.; Wang, X.; Chen, F.; Li, J.; Wu, J.; Sun, Y.; Zhang, Y.; Deng, T.; Jiang, S.; Zhou, X.; et al. Soil organic matter enhances aboveground biomass in alpine grassland under drought. *Geoderma* **2023**, *433*, 116430. [CrossRef]

applied sciences

MDPI

Article

Time Series Prediction Model of Landslide Displacement Using Mean-Based Low-Rank Autoregressive Tensor Completion

Chenhui Wang [1,2,*] and Yijiu Zhao [1]

1 School of Automation Engineering, University of Electronic Science and Technology of China, Chengdu 611731, China
2 Center for Hydrogeology and Environmental Geology Survey, China Geological Survey, Baoding 071051, China
* Correspondence: wangchenhui@mail.cgs.gov.cn

Abstract: Landslide displacement prediction is a challenging research task that can help to reduce the occurrence of landslide disasters. The frequent occurrence of extreme weather increases the probability of landslides, and the subsequent increase in the superimposed economic development level exacerbates disaster losses, emphasizing the importance of landslide prediction. The collection of landslide monitoring data is the foundation of landslide displacement prediction, but the lack of various data severely limits the effectiveness of the landslide monitoring system. To address the issue of missing data during the landslide monitoring process, this paper proposes a time series prediction model of landslide displacement using mean-based low-rank autoregressive tensor completion (MLATC). Firstly, the reasons for the missing data of landslide displacement are analyzed, and the corresponding dataset of missing data is designed. Then, according to the characteristics and internal correlation of landslide displacement monitoring data, the establishment process of mean-based low-rank tensor completion prediction model is introduced. Finally, the proposed method is used to complete and predict the missing data for the random missing and non-random missing landslide displacement. The results show that the data completion and prediction results of the model are essentially consistent with the original displacement monitoring data of the landslide, and the accuracy and precision are relatively high. It shows that the model has good landslide displacement completion and prediction effects, which can provide a certain reference value for the missing data processing and landslide displacement prediction.

Keywords: time series; missing data; tensor completion; autoregressive norm; displacement prediction

Citation: Wang, C.; Zhao, Y. Time Series Prediction Model of Landslide Displacement Using Mean-Based Low-Rank Autoregressive Tensor Completion. *Appl. Sci.* **2023**, *13*, 5214. https://doi.org/10.3390/app13085214

Academic Editors: Daniel Dias, Yuzhu Wang, Jinrong Jiang and Yangang Wang

Received: 6 February 2023
Revised: 7 April 2023
Accepted: 19 April 2023
Published: 21 April 2023

1. Introduction

Landslides are one of the many common natural disasters in the world [1,2]. A frequent occurrence of extreme weather increases the likelihood of landslides. Rapid economic and social development further aggravates the loss of landslide disasters [3,4]. Therefore, accurate landslide displacement prediction is becoming increasingly important in order to prevent and mitigate the damage caused by landslides [5]. In the process of automated real-time monitoring of landslides, data acquisition and transmission are generally performed using different types of sensors and other electronic devices [6]. Automated monitoring equipment is always in the open-air environment, and most of them inevitably suffer from tear, aging, power loss and other phenomena, all of which can lead to missing monitoring data. In addition, most landslide disasters are located in a relatively harsh geological environment, such as heavy rainfall, hail, dense fog, electromagnetic interference, etc., and the installation and deployment of geological hazard monitoring equipment in open fields will inevitably be affected by the abovementioned harsh environment. Randomness or prolonged interruptions in the operation of the monitoring device can cause the monitoring device to fail to properly send monitoring data to the server, which leads to the problem of missing monitoring time series in the server. Time series forecasting is a valid basis for making accurate discriminations. To enhance the accuracy of landslide prediction, it is necessary to construct a corresponding accurate data completion method.

For the problem of completing and predicting missing landslide data, most traditional time series models have focused on models such as regression analysis and exponential smoothing. The problem of incompleteness for missing time series data can be broadly classified as either deletion or padding. Data deletion is used as anomalous data to remove some objective abnormal monitoring data, which is primarily used for anomaly detection and feature analysis, while filling is utilized to find the long-term time series change pattern of monitoring data and to supplement the missing monitoring data. The main methods include missing value filling algorithm based on nearest neighbor method, cyclic neural network, random forest and matrix decomposition, but more data are required for machine learning training [7,8]. Statistical filling is more effective for data series with less dimensions that can establish a maximum model, provided that the relationship between missing eigenvalues and existing eigenvalues can be established through observation. The main methods of machine learning include missing value filling algorithm based on the nearest neighbor method, cyclic neural network, and matrix decomposition. Matrix decomposition can effectively explore the correlation between different time series for different dimensions of long time series problems. The matrix decomposition method is used to learn the overall characteristics of the time series matrix, which can be used to approximate the matrix with time characteristics in low-rank, and then complete the missing data [9].

Large-scale time series data are always accompanied by the missing problem. Therefore, the tensor completion method has been introduced into this field to complement the traditional data completion method based on probability and statistics [10–12]. The data completion scheme based on simple quantitative statistics has a relatively simple and efficient processing effect for small datasets and simple regression models, but it is not feasible for massive data in the era of data explosion. Modern research not only has many kinds of data variables and long time series, but also requires fast processing speed, high universality and portability. Multiple variables and long time series can better describe the complex causal relationship between each other [13,14]. Now, neural network technology is often used to deal with the above situations and delete the missing data in order to form a complex intelligent model, but it is not the best solution for missing data because deletion may strengthen or weaken the connection degree of a causal relationship. Based on this, scholars have explored the application of tensor decomposition technology to data completion in multivariate long time series, trying to improve the resolution speed and data missing problems.

Recent studies have found that low-rank matrices have certain advantages in the analysis of multivariate long time series data [11,12], including the sequence tensor completion method [15]. The sequence tensor complement restores the potential tensor from the sampling structure of the time series, allocates the position of the missing items as needed, seamlessly integrates the future value of the time series into the framework of the missing data and improves the data completion accuracy [15]. The low-rank matrix completion method performs singular spectrum analysis and singular value decomposition on the time series in order to complete the low-rank completion of the missing data of the time series, although its calculation is large [16]. Therefore, by adding a time dimension to transform the status time series into a high-dimensional tensor, the cost of computing complexity is better solved. This is also in line with the law of human activities, both short and long term activities, so there are studies using tensors (sensors $\times$ 1 day $\times$ 24 h) which indicate the above activity mode [17,18]. The dependency between sensors is preserved, providing a new feasible scheme for capturing local and global time patterns [16]. More scholars have combined the autoregressive moving average model with the tensor model to propose the low-rank matrix autoregressive tensor completion model and have achieved good results in the completion and prediction of financial time series data [19].

The mean-based low-rank autoregressive tensor completion mainly includes completing low-rank matrix decomposition/tensor completion and constructing time series autoregressive models, as well as processing the missing data with the neighboring data mean instead of zero before the operation. The low-rank matrix completion model uses the underlying low-rank structure to recover incomplete matrices (assuming that the long-term landslide data sequence is incomplete) [20]. Considering that the deformation displacement of landslide has a great correlation with the previous deformation, the autoregressive model is constructed to represent the deformation law of landslide displacement with time. The autoregressive regularizer is introduced in the low-rank matrix decomposition to characterize the temporal dynamics in landslide displacement deformation, and the learned autoregressive regularizer is implemented to predict the temporal factor matrix, thus realizing the landslide displacement monitoring data completion and predictive modeling [12].

The purpose of this study is to establish a new method for completing and predicting landslide displacement data based on MLATC. In this paper, the causes of data loss of landslide displacement are analyzed. Taking the Shuizhuyuan landslide in the Three Gorges Reservoir area as an example, the data completion and prediction algorithm are designed by using MLATC. Then, the landslide displacement data are divided into training set and test set, and the random missing and non-random missing are selected for corresponding data completion and prediction. The designed model can achieve an accurate completion and prediction of landslide displacement. Finally, a comparative analysis with existing models verifies the effectiveness of the model.

2. Theory and Method

2.1. Reasons and Analysis of Missing Data

2.1.1. Reasons of Missing Data

The reasons for the missing of landslide monitoring data are complex, which may be caused by the process of data collection, transmission, storage and analysis. Time series with missing data typically share the same characteristics, such as noisy, incomplete and abnormally abrupt data in the time series. Missing data in the time series can have a significant impact on future landslide data analysis, monitoring and early warning. The data should thus be preprocessed in the process of data analysis, i.e., data cleaning, and the processing of missing data is one of the key elements in data cleaning.

Data deficiencies in landslide monitoring time series are broadly characterized as follows. (1) Long time span and large amount of data: Landslide monitoring requires the data collection of the different factors affecting landslide displacement and deformation, and the continuous extension of monitoring time leads to the increasing amount of data. (2) Randomness: The data acquisition and transmission in the landslide monitoring system are completed automatically by the field equipment, and the components and modules used for acquisition and transmission are all electronic components. It is highly susceptible to rainfall, hail, electromagnetic interference and other natural factors, and the above natural phenomena have randomness. (3) Spatial correlation: The mechanism of landslide disasters is complex. Corresponding monitoring devices will be deployed in different areas of the landslide, and there will be some correlation between different sensors. Therefore, missing data completion needs to consider the spatial correlation among the sensors in the time series.

Due to the diverse properties of the landslide time series and the different types of monitoring data, there are missing data in the time series for a variety of causes, which can be roughly summarized as follows. (1) Data are not available: The landslide monitoring process is affected by the harsh natural environment. Problems with the power supply of the equipment and the transient failure of the sensors can lead to a certain period of time or a certain moment of failure to complete the data acquisition, thus leading to the lack of data. (2) Data transmission failure: The transmission of landslide monitoring data mainly relies on a wireless communication network; when the wireless signal is interrupted, the data for a certain period of time or a certain moment cannot be transmitted to the background monitoring center. (3) Human interference: When technicians perform data preprocessing operations, the relevant data may be transmitted incorrectly for various reasons, which may also lead to data loss.

2.1.2. Types of Missing Data

According to the existing studies, different patterns of missing data occur in different time series. Considering the actual situation of landslide monitoring, the types of missing data in landslide monitoring time series can be usually classified into the following two types.

(1) Random missing (RM): Randomization can occur at any time during the monitoring process. Such data loss has no regularity or symptoms. The time of data loss is random and occasional, as shown in Figure 1.

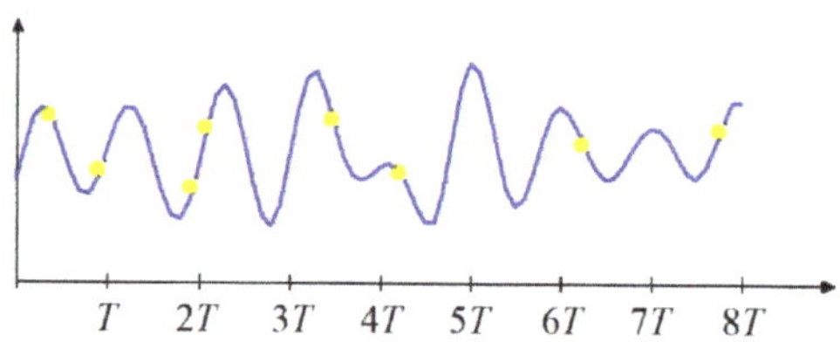

Figure 1. Random missing. The yellow dots represent missing data.

(2) Non-random missing (NM): In the monitoring process, regular or periodic phenomena such as cloudy weather and signal interruptions may occur with the same probability in a specific period of time, and such missing data has a certain regularity, as shown in Figure 2.

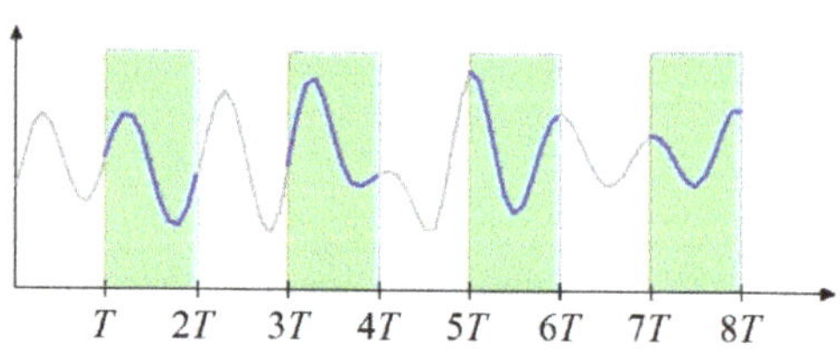

Figure 2. Non-random missing. The curves in the green area represent missing data.

2.2. *Construction of Low-Rank Tensor Completion Model*

2.2.1. Tensor Composition

In the landslide monitoring system, the monitoring data of each on-site monitoring sensor is collected through the intelligent terminal, and then the on-site monitoring data are transmitted to the background data monitoring center. Considering that the collected monitoring data include date, time, displacement and other information, it can be considered that the acquired landslide displacement time series is high-dimensional. For the landslide time series, this paper constructs a three-dimensional missing tensor $\mathcal{T} \in R^{l_1,l_2,l_3}$, where l_1 is the number of on-site sensors for landslide, l_2 is days collected, and l_3 is the frequency of sampling every day. Therefore, the tensor represents the displacement monitoring value of each sensor at the corresponding time. This paper aims to identify the missing data position in tensor $\mathcal{T}$ and complete data completion and displacement prediction, respectively.

2.2.2. Construction of Low-Rank Tensor Completion Model

The low-rank matrix completion (LRMC) model uses the low-rank structure at the bottom to restore the incomplete matrix [21], defining the rank of the tensor $\mathcal{X} \in R^{l_1,l_2,l_3}$ as rank($\mathcal{X}$) and the objective function of low-rank tensor completion as:

$$\begin{cases} \min\limits_{\mathcal{X}} \text{rank}(\mathcal{X}) \\ s.t. \mathcal{X}_\Omega = \mathcal{T}_\Omega \end{cases} \tag{1}$$

where $\mathcal{X} \in R^{l_1,l_2,l_3}$ is a tensor to be solved; $\mathcal{T} \in R^{l_1,l_2,l_3}$ is a missing tensor; Ω is the observed landslide data in T. Then, the tensor kernel norm is replaced by the rank minimum [22], and the concept of tensor kernel norm is defined:

$$\| \mathcal{X} \|_* = \sum_{i=1}^{3} \alpha_i \| X_{(i)} \|_* \tag{2}$$

where α_i is the weighting factor and $X_{(i)}$ is the matrix expanded along the i-th mode for tensor $\mathcal{X}$. The three pattern expansion matrices of the data tensor $\mathcal{X} \in R^{l_1,l_2,l_3}$ are shown in Formula (3):

$$\begin{cases} X_{(1)} \in R^{l_1 \times (l_2 \times l_3)} \\ X_{(2)} \in R^{l_2 \times (l_1 \times l_3)} \\ X_{(1)} \in R^{l_3 \times (l_1 \times l_2)} \end{cases} \tag{3}$$

where $X_{(1)}, X_{(2)}, X_{(3)}$ are the module expansion matrix for the staggered sampling of three types of expansion patterns, respectively.

The features of different orders between data are fused with each other through three module expansion matrices, effectively ensuring the high-precision completion of data [23]. Below is the singular value decomposition of each pattern expansion matrix [20]:

$$\begin{cases} X_{(1)} = U_{l_1 \times l_1} \sum\limits_{l_1 \times l_2 l_3}^{(1)} {M^T}_{l_2 l_3 \times l_2 l_3} \\ X_{(2)} = V_{l_2 \times l_2} \sum\limits_{l_2 \times l_1 l_3}^{(2)} {L^T}_{l_1 l_3 \times l_1 l_3} \\ X_{(3)} = W_{l_3 \times l_3} \sum\limits_{l_3 \times l_1 l_2}^{(3)} {K^T}_{l_1 l_2 \times l_1 l_2} \end{cases} \tag{4}$$

where U, V and W are left singular matrices; M, L and K are right singular matrices; and Σ is a diagonal matrix. It is defined as:

$$\sum{}^{(q)} = diag\left(\sigma_1{}^{(q)}, \sigma_2{}^{(q)}, \cdots, \sigma_{n_q}{}^{(q)}\right), q = 1,2,3 \tag{5}$$

where n_q is the total number of singular values of the q-th module expansion matrix. Based on the definition of tensor kernel norm, the objective function is:

$$\begin{cases} \min\limits_{\mathcal{X}} \sum\limits_{i=1}^{3} \alpha_i ||X_{(i)}|| \\ s.t. \mathcal{X}_\Omega = \mathcal{T}_\Omega \end{cases} \tag{6}$$

where the matrix core norm is defined as [24]:

$$\| X \|_* = \sum_{k=1}^{j} \sigma_k(X) \tag{7}$$

where $\|\cdot\|_*$ is the kernel norm of X; j is the rank of matrix X; σ_k is the k-th singular value of the matrix X in order of size. Because of the existence of interdependent matrix kernel norm terms, the objective function of low-rank tensor completion is difficult to be solved by ordinary methods. Alternating direction method of multipliers (ADMM) [23,25] is a more widely used optimization method for constrained problems in machine learning. It is an extension of the augmented Lagrange method. ADMM algorithm provides a framework for solving constrained optimization problems with linear equations, which is convenient for disassembling the original optimization problem into several relatively solvable sub-optimization problems for iterative solutions. With continuous research, the low-rank tensor completion method is further developed. The purpose of reducing the computational complexity is to use QR decomposition instead of singular value decomposition. The decomposed low-rank tensor completion reduces the time complexity of the low-rank tensor completion algorithm. The nonlinear set CG algorithm based on Riemannian manifold reduces the decomposition of large-scale singular values in low-rank tensor completion. The tensor completion algorithm based on the Douglas–Rachford separation technology further considers the existence of the noise of the source data [25], which greatly enhances the robustness of the model. In addition, low-rank tensor completion also has a series of characteristics such as fast convergence speed and high computational accuracy.

2.3. *Construction of MLATC*

The model first transforms time series matrix into a tensor, which is recorded as sign $\mathcal{Q}(\bullet)$, such as $\mathcal{Q}(Y)$, transforming matrix $Y \in \mathrm{R}^{M\times N}$ into a third-order tensor of size $M \times N \times J$ [12].

2.3.1. Zero-Valued Low-Rank Autoregressive Tensor Completion Core Model

$$\min_{\mathcal{X},\mathcal{Z},A} \| \mathcal{X} \|_* + \lambda \| \mathcal{Z} \|_{A,\mathcal{X}} \quad s.t. \begin{cases} \mathcal{X} = \mathcal{Q}(Z) \\ \mathcal{P}_\Omega(Z) = \mathcal{P}_\Omega(Y) \end{cases} \tag{8}$$

where Y is the input observation matrix, and the matrix size of Z and Y is the same. $\| \mathcal{X} \|_*$ represents the kernel norm of the tensor $\mathcal{X}$, and the calculation method is $\| \mathcal{X} \|_* = \sum_k a_k \| X_{(k)} \|_*$; $\| X \|_* = \sum_{i=1}^{\min\{M,N\}} \sigma_i$. σ_i is the i-th largest singular value of matrix X; A is the variable to be estimated; λ is a control objective function to balance the weight parameters of the first and second items. $\| \mathcal{Z} \|_{A\mathcal{H}}$ represents the autoregressive norm of matrix Z, whose value is determined by Equation (9).

$$\| \mathcal{Z} \|_{A,\mathcal{H}} = \sum_{m,t} \left(z_{m,t} - \sum_i a_{m,t} z_{m,t-h_i} \right)^2 \tag{9}$$

$\mathcal{P}_\Omega(Z)$ represents the orthogonal projection of matrix Z in the field Ω, and the calculation formula is:

$$[\mathcal{P}_\Omega(Z)]_{m,n} = \begin{cases} z_{m,n} & if(m,n) \in \Omega \\ 0 & if(m,n) \notin \Omega \end{cases} \tag{10}$$

2.3.2. Mean-Based Low-Rank Autoregressive Tensor Completion

The time series matrix $Y \in \mathrm{R}^{M\times(N=I\times J)}$ with partially missing data can be compressed into a tensor $Y \in \mathrm{R}^{M\times I\times J}$, thus transforming the matrix completion problem into a tensor completion problem. The traditional tensor completion method assigns the missing data to zero. Considering the time accumulation and continuity of landslide displacement and deformation, the average value of the data before and after the missing data are taken instead of assigning the value to zero when completing the landslide displacement data.

The specific difference is that the projection of matrix Z in the domain Ω is further required to obtain a useful non-orthogonal projection:

$$[\mathcal{P}_\Omega(Z')]_{m,n}{}' = \begin{cases} [\mathcal{P}_\Omega(Z)]_{m,n} & if(m,n) \in \Omega \\ \frac{1}{2t}\sum\limits_{i=n-t}^{n+t} [\mathcal{P}_\Omega(Z)]_{m,i} & if(m,n) \notin \Omega \end{cases} \tag{11}$$

where t is a hyperparameter representing the time sequence length of taking the mean value, in this case, 10 is taken. The matrix Z' conforming to such constraints is called a close orthogonal matrix of Y in the field Ω (recorded as $Z' \overset{\perp}{\sim} Y_\Omega$). For ease of understanding, Z' is still noted as Z.

MLATC is a simple linear combination of LRMC and vector auto-regressive (VAR) models. The first term of the objective function is the LRMC objective function, and the second term is the VAR objective function. The Lagrange function in matrix form is constructed according to the optimization problem:

$$L(\mathcal{X}, Z, A) = \sum_k a_k \| \mathcal{X}_{k(k)} \|_* + \lambda \| \mathcal{Z} \|_{A,\mathcal{H}} + \left\langle \mathcal{Q}^{-1}(\mathcal{X}) - Z, \mathcal{Q}^{-1}(\tau_k) \right\rangle + \frac{\rho}{2} \| \mathcal{Q}^{-1}(\mathcal{X}) - Z \|_F^2 \tag{12}$$

Then, Question (12) is transformed into $argmin_{\mathcal{X}_{k(k)}, Z, A} L(\mathcal{X}, Z, A, \tau_k)$, where X and Z are variables, and τ_k is the parameters to be learned. When solving the tensor $\mathcal{X}_k$, $\frac{\partial L(\mathcal{X},Z,A)}{\partial \mathcal{X}} = 0$ should hold. According to Formula (9), since $\frac{\partial \|Z\|_{A,\mathcal{H}}}{\partial \mathcal{X}} = 0$, $argmin_{\mathcal{X}} L(\mathcal{X}, Z, A) \Leftrightarrow argmin_{\mathcal{X}} L(\mathcal{X}, Z, A) - \| Z \|_{A,\mathcal{H}}$. Thus, $\mathcal{X}_k^{l+1}$ can be resolved by Formula (13):

$$\mathcal{X}_k^{l+1} = arg \min_{\mathcal{X}} a_k \| \mathcal{X}_{(k)} \|_* + \frac{\rho}{2} \| \mathcal{Q}^{-1}(\mathcal{X}) - Z^l \|_F^2 + \left\langle \mathcal{Q}^{-1}(\mathcal{X}) - Z^l, \mathcal{Q}^{-1}\left(\tau_k^l\right) \right\rangle \tag{13}$$

In Equation (13), the number l of iterations of the algorithm is expressed, and the VAR model is used to solve the matrix Z [12,21,26].

The MLATC achieves a more accurate completion of the original data than the zero-filled algorithm, and then the prediction of landslide displacement data can be completed on this basis.

3. Case Study

3.1. Experimental Dataset

The displacement monitoring data of the Shuizhuyuan landslide in the Three Gorges Reservoir area is selected as the time series' experimental dataset of this study. This time series is collected from seven GNSS monitoring sensors at the Shuizhuyuan landslide. The time interval is from 15 July 2017 to 1 December 2021, with a total of 1600 days. The data monitoring cycle is to obtain one monitoring data at the site every day, so the tensor size of the constructed dataset is 7 × 1600 × 1 (sensors × time points × day) tensor structure. The monitoring data are shown in Figure 3.

3.2. Missing Data Processing for Time Series

In order to objectively describe the validity of the prediction model, two different data missing processing methods, random and non-random, are used to carry out certain data missing processes on the original time series dataset of landslides by combining the characteristics and main types of missing landslide monitoring data.

Random missing (RM): Random missing indicates that the landslide monitoring equipment has sporadic and random data loss in the operation process. The whole time series is divided according to the proportion of random missing data, with 5%, 10%, 20% and 40% set, respectively, representing different scales and levels of data loss of monitoring equipment.

Non-random missing (NM): Non-random missing means that the landslide monitoring equipment has regular equipment failure or regular network interruption during operation.

The data missing is also simulated according to 5%, 10%, 20% and 40% and compared with random missing.

The original spatiotemporal landslide displacement dataset is artificially treated with missing data, and the effectiveness of different models in completing and predicting the missing data can be objectively evaluated by two different treatments, random as well as non-random. For landslide disasters with relatively complex genetic mechanisms, there must be certain trend characteristics and a correlation between different sensors on the landslide. Therefore, better data recovery and prediction for missing datasets after non-random missing processing is an important indicator of this model.

Shuizhuyuan landslide is currently in the mean-slow deformation, and the first 1300 days are selected as the training set of the time series, and the last 300 days are selected as the test set. The MAPE $=\sum_{i=1}^{n}\left|\frac{\hat{x}_i - x_i}{x_i}\right| \times \frac{100\%}{n}$ and RMSE $=\sqrt{\frac{1}{n}\sum_{i=1}^{n}(\hat{x}_i - x_i)^2}$ are used to test the accuracy and precision of the algorithm model in terms of prediction. Moreover, the MLATC model is compared with the high-accuracy low-rank tensor completion (HALRTC) and temporal regularized matrix factorization (TRMF) methods. The completion and prediction results with a smaller MAPE and RMSE are considered to be better.

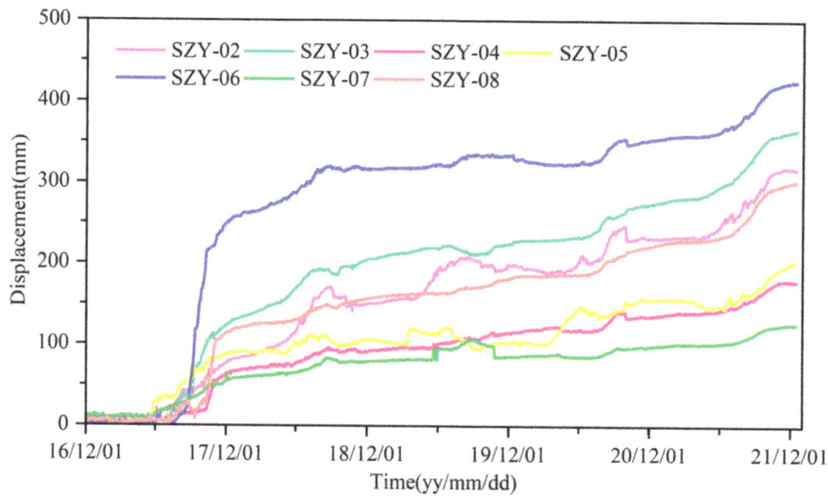

Figure 3. Landslide displacement monitoring data.

3.3. Data Completion and Analysis

The experimental results of four sensors numbered SZY-02, SZY-03, SZY-06 and SZY-08 are used as examples to introduce the effect of landslide monitoring data completion. After NM40% processing of the dataset of Shuizhuyuan landslide, the data completion effect of the completed training set of SZY-08 is shown in Figure 4. In Figure 4, the blue curve represents the measured data, the yellow dots represent the missing data, the red curve represents the recovered data, X-axis 0–260, 260–520, 520–780, 780–1040, 1040–1300 represents data recorded for each day and Y-axis is the displacement data. Considering the large amount of data for each sensor, only SZY-08 shows the complementary results for the entire time series, and the other sensors are the complementary results for the first 260 days. Figure 5 shows the complementary effect of the other sensors for the first 260 days.

As can be seen in Figures 4 and 5, in the absence of raw landslide displacement data for many consecutive days, the data have been completely processed into missing data, which is challenging for tensor completion. Missing data for several consecutive days has lost the correlation and trend characteristic information between displacement data. The prediction results show that the MLATC model still achieves a very good complementary effect, which is essentially consistent with the original displacement data and achieves a good effective data recovery in some completely missing data days. This is very helpful to analyze the deformation law and deformation trend of the landslide, and also to provide data reference for understanding the relationship between different deformation areas of the landslide.

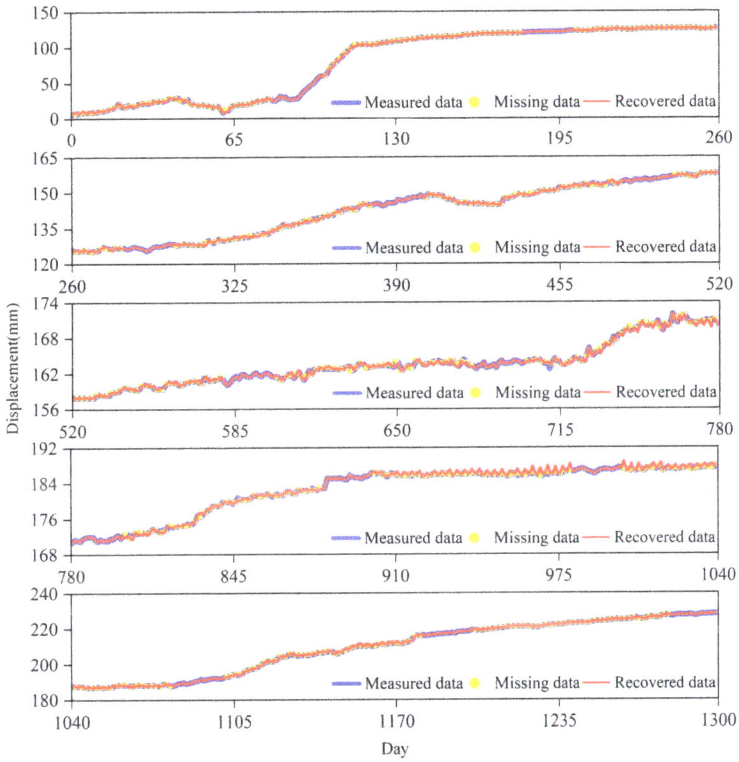

Figure 4. Completion of displacement data of SZY-08 (NM40%). The blue curve represents the measured data, the yellow dots represent the missing data and the red curve represents the recovered data.

Table 1 shows the evaluation metrics for data completion in the non-random missing case. The results in Table 1 show that the MLATC model has the best MAPE and RMSE for four different scenarios of NM5%, NM10%, NM20% and NM40%, indicating that the MLATC model has a greater improvement in data recovery performance compared to the HALRTC and TRMF models.

Table 1. Evaluation metrics for data completion in the NM case (MAPE/RMSE).

Model	TRMF	HALRTC	MLATC
NM 5%	11.8542/14.0631	1.3161/3.8315	0.9066/0.9196
NM 10%	10.2681/16.5992	1.1953/3.6270	0.7770/0.8958
NM 20%	11.7050/18.1097	11.7976/25.2256	0.7708/0.9616
NM 40%	13.8412/25.9528	15.0777/38.6958	0.7928/1.0070

From Table 1, it can be concluded that the MLATC model is better than the HALRTC and TRMF models in terms of completion effect, indicating that the tensor autoregressive kernel parametrization can effectively replace the rank function, which enables the low-rank tensor completion model to obtain a more accurate completion effect. In addition, the MLATC model introduces an autoregressive norm on the basis of the HALRTC model, which can make full use of the structural correlation and local trend feature information between the high-dimensional multivariate time series data, as well as more clearly correlate the multivariate time series data, which also proves that the addition of the autoregressive norm can further improve the complementary performance and accuracy of the model in the low-rank tensor complementary structure. Since the tensor structure well maintains the structural information of the spatiotemporal data in the time dimension and exploits the correlation between the daily displacements of different sensors in the time series, the tensor completion results of the MLATC model are better than those of TRMF in the matrix form. The experimental results with different missing ratios simultaneously verify this inference and confirm that effective completion and rolling prediction of displacement monitoring data can be achieved by using global time series datasets. After analyzing the reasons, the TRMF model only analyzes and calculates the matrix structure and does not use the time domain smoothness of multivariate time series and the potential correlation information between the series in the time and space dimensions. MLATC and HALRTC both use the tensor structure to complete the time series prediction through the method of quantitative completion. In particular, the MLATC model introduces the autoregressive norm, which not only preserves the structural information of the original time series through the tensor structure, but also makes full use of the correlation and trend characteristic information between the time series.

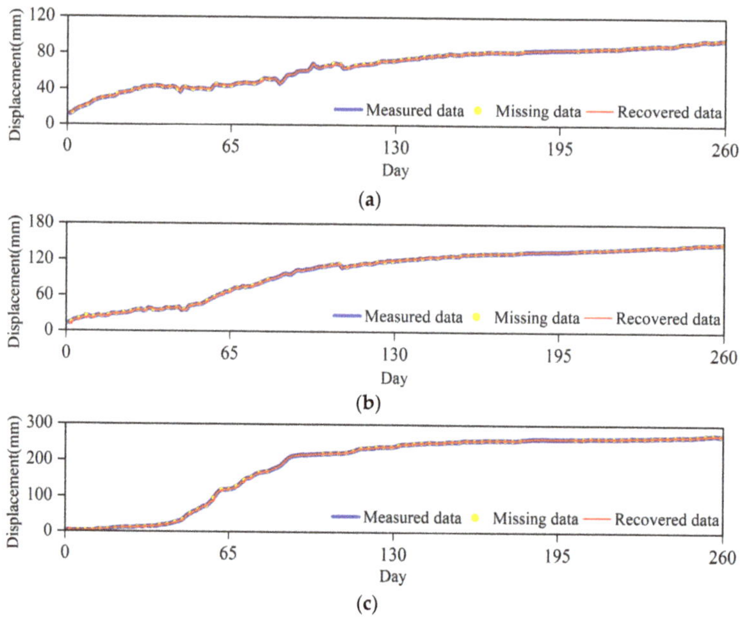

Figure 5. Completion of displacement data (NM40%). (**a**) Point SZY-02; (**b**) point SZY-03; (**c**) point SZY-06. The blue curve represents the measured data, the yellow dots represent the missing data and the red curve represents the recovered data.

Similarly, the experimental results of the four sensors numbered SZY-08, SZY-02, SZY-03 and SZY-06 are shown in Figures 6 and 7 after RM 40% processing of the dataset of the Shuizhuyuan landslide; the data completion performance are shown in Table 2. The prediction model has a very good recovery effect on missing data and shows a better data fit in terms of time series trend prediction.

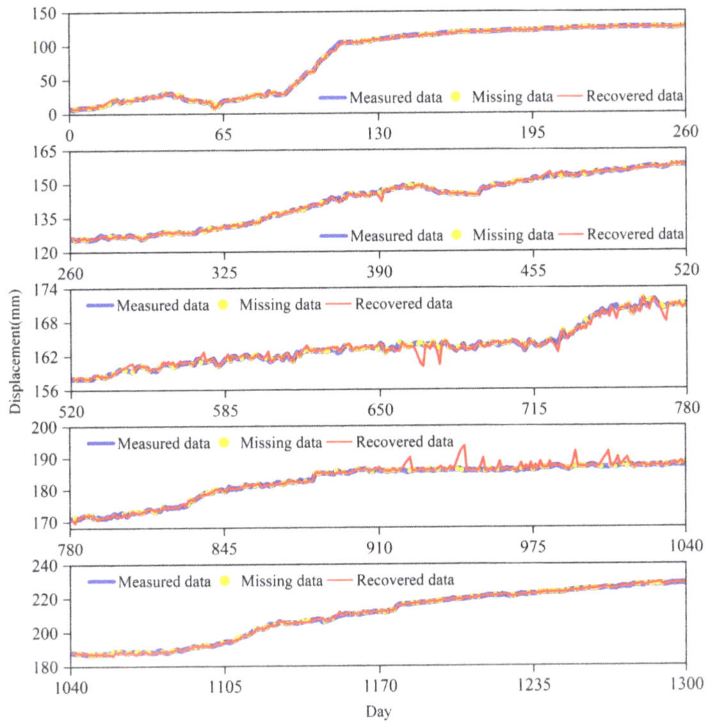

Figure 6. Completion of the displacement data of SZY-08 (RM40%). The blue curve represents the measured data, the yellow dots represent the missing data and the red curve represents the recovered data.

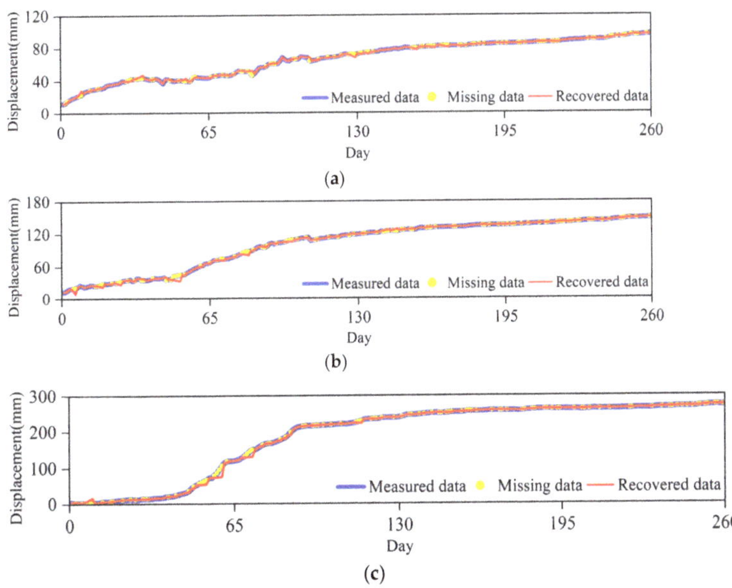

Figure 7. Completion of displacement data (RM40%). (a) Point SZY-02; (b) point SZY-03; (c) point SZY-06. The blue curve represents the measured data, the yellow dots represent the missing data and the red curve represents the recovered data.

Table 2. Evaluation metrics for data completion in the RM case (MAPE/RMSE).

Model	TRMF	HALRTC	MLATC
RM 5%	11.8535/14.0634	0.8087/2.7628	0.7676/0.9880
RM 10%	10.2782/16.5925	0.9729/3.0545	0.7263/1.0339
RM 20%	11.7037/18.1142	11.7976/25.2256	0.9494/1.7648
RM 40%	13.8347/25.9556	15.0777/38.6958	1.4817/1.8889

3.4. Data Prediction and Analysis

In the case of random missing, the time series of the Shuizhuyuan landslide is predicted for 300 days after data missing processing is performed with a missing ratio of 40%. The time series of four sensors numbered SZY-02, SZY-03, SZY-06 and SZY-08 are analyzed separately, and the prediction results are shown in Figure 8. The analysis shows that the prediction results are ultimately consistent with the original displacement data in the case of the 40% missing data ratio. The MLATC model realizes well the deformation trend feature fitting of displacement, and effectively predicts the displacement data.

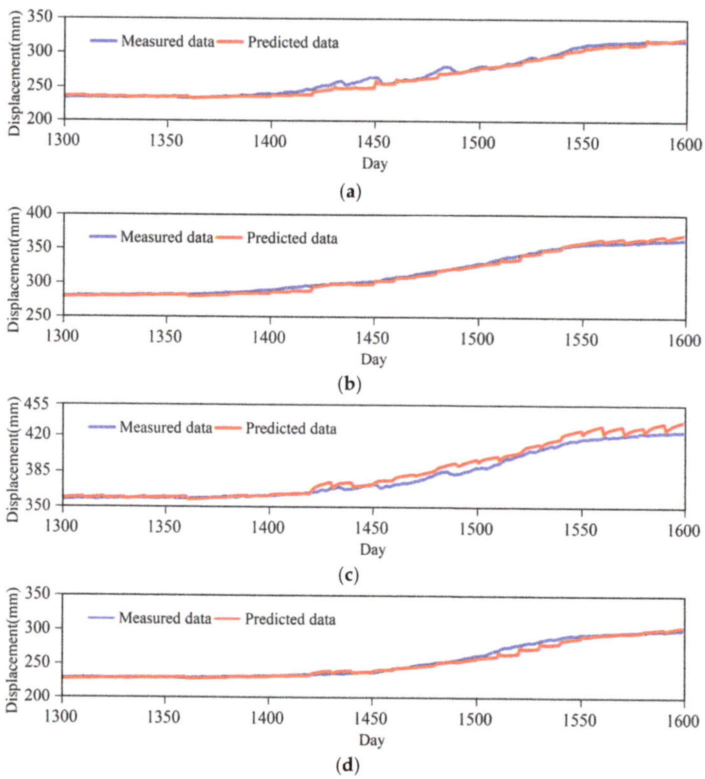

Figure 8. Prediction of displacement data (RM40%). (**a**) Point SZY-02; (**b**) point SZY-03; (**c**) point SZY-06; (**d**) point SZY-08. The blue curve represents the measured data and the red curve represents the predicted data.

Similarly, after processing the dataset of the Shuizhuyuan landslide with 40% non-random missing, the data completion effect is shown in Figure 9. The prediction results of MLATC model are also ultimately close to the original displacement data. Although there are some fluctuations in the prediction data, the overall effective fitting of displacement is still achieved.

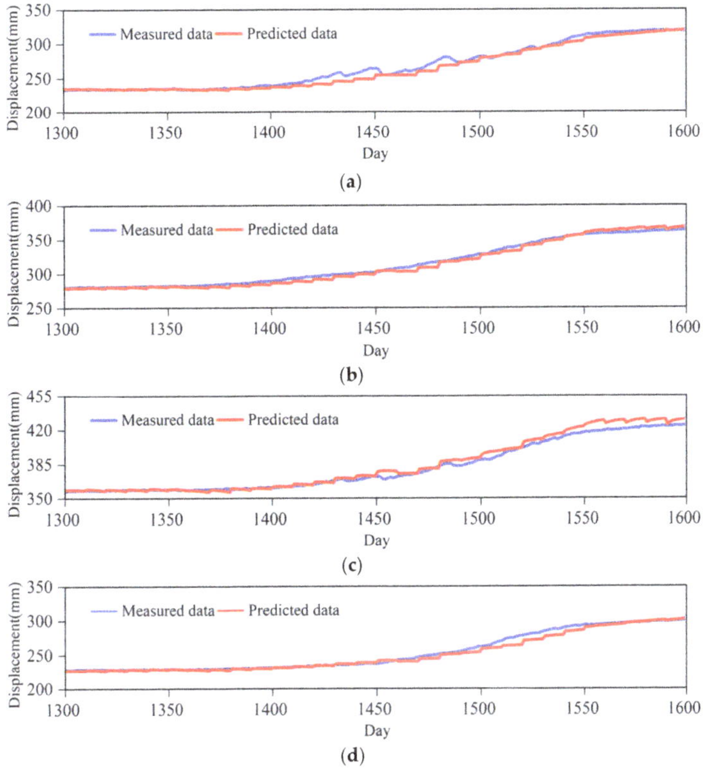

Figure 9. Prediction of displacement data (NM40%). (**a**) Point SZY-02; (**b**) point SZY-03; (**c**) point SZY-06; (**d**) point SZY-08. The blue curve represents the measured data and the red curve represents the predicted data.

The prediction effect of MLATC model under random and non-random missing is shown in Tables 3 and 4.

Table 3. Evaluation metrics for data prediction in the RM case (MAPE/RMSE).

Model	TRMF	HALRTC	MLATC
RM5%	39.6023/102.415	15.2062/39.8024	1.1084/3.6774
RM10%	41.1064/106.681	14.8872/39.0352	1.0792/3.6587
RM20%	50.2660/127.588	14.8241/39.1718	1.1120/3.7032
RM40%	83.7915/213.040	13.8827/37.1245	1.1012/3.6665

Table 4. Evaluation metrics for data prediction in the NM case (MAPE/RMSE).

Model	TRMF	HALRTC	MLATC
NM5%	39.4527/102.047	14.9328/38.9776	1.1079/3.6676
NM10%	42.7754/110.746	14.2427/37.2552	1.1037/3.6588
NM20%	49.9770/126.921	12.6523/33.3975	1.1099/3.6785
NM40%	83.7004/212.867	11.5458/30.6736	1.1694/3.8755

By analyzing the prediction results of each model, it can be proven that the prediction accuracy of the time series model is effectively improved after converting the original time series into a tensor structure in this method. The prediction accuracy of both HALRTC model and MLATC model has improved significantly over the prediction accuracy of

TRMF. This indicates that although the TRMF model captures the global consistency of the time series, the model mainly acts on the latent layer matrix data, which may cause the local trend feature information between different sensors to be ignored, thus affecting the prediction accuracy. Due to the intervention of the autoregressive norm, the prediction of HALRTC and MLATC models is significantly better than that of TRMF model, and the MAPE and RMSE accuracy of the models are higher. This demonstrates that the spatiotemporal monitoring data between displacement sensors in different areas of the landslide are utilized under the framework of tensor complementary structure, and the deformation law characteristics of short-term correlation and long-term deformation consistency in the inherent deformation process of the landslide are fully considered.

4. Discussion

Affected by the complex environment in the field, there will be missing data in the process of landslide monitoring, which will affect the accurate analysis of landslide displacement. In landslide disaster monitoring, landslide displacement deformation is relatively complex, and the displacement changes of each deformation area on the landslide are not the same. From the analysis of landslide local deformation characteristics, there is a certain correlation between the daily displacement deformation of different monitoring points. In addition, landslide displacement is a continuous cumulative process, and the landslide displacement occurring in the preceding period will have a certain influence on the landslide displacement occurring subsequently. Therefore, in order to better realize landslide data completion and prediction, a novel model based on MLATC method is proposed, and the RM and NM cases are designed, respectively, so as to verify the validity and reliability of the designed model. In addition, from the construction of the model, the initial time series matrix is converted into a third-order tensor structure. Under the assumption that the time series data satisfy approximate low-rankness, the time series completion and prediction problem is transformed into low-rank tensor completion and prediction. The tensor-based completion method fully considers the time series of landslide displacement data, which not only preserves the correlation between different displacements, but also better fits the landslide displacement deformation characteristics, making the displacement completion more accurate.

To verify the effectiveness of the algorithm, a comparative analysis with the existing RTMF and HALRTC models was performed. The MAPE and RMSE of the MLATC model are 0.9066, 0.9196 and 0.7676, 0.9880 for the NM5% and RM5% data completion, respectively. Similarly, the MAPE and RMSE for NM5% and RM5% data prediction are 1.1079, 3.6676 and 1.1084, 3.6774, respectively. The analysis results show that under the conditions of 5%, 10%, 20% and 40% missing data, the data completion and prediction effects of the MLATC model are better than those of other models, which also confirms the significant data completion and prediction effects of the MLATC model. In this study, the completion and prediction model were constructed by considering the intrinsic correlation between landslide displacement data and using the low-rankness of the completion tensor. Data completion is an iterative calculation of the landslide displacement time series based on the entire original data. In order to verify the effectiveness of the data completion model, random and non-random data missing cases are designed, and time series prediction is based on the predicted experimental data as the missing value, which also belongs to a special case of missing data. By default, this part of the experimental data is not involved in iterative calculations. Therefore, the data prediction for time series is the same as the method used for data completion. The model is implemented in a rolling prediction method with cyclic and iterative computation, which leads to a loss in the amount of data in the prediction case. The MAPE and RMSE values of NM and RM also show that the prediction effect of the MLATC model is lower than that of data completion, but both can achieve satisfactory results, which are more in line with the actual needs of landslide monitoring.

5. Conclusions

A novel method for landslide displacement data completion and prediction based on MLATC model is proposed for missing landslide displacement data and time series prediction. The tensor structure in the MLATC model well perpetuates the structural information of landslide spatiotemporal data in the time dimension, makes full use of the correlation of landslide displacement time series in different time scales, and solves the problem of serious data loss caused by the destruction of the original matrix structure. The data completion and prediction models are implemented in a tensor structure framework, combining VAR models and autoregressive paradigms, and different proportions of random and non-random missing cases are selected for experimental analysis. The experimental results of the Shuizhuyuan landslide prove that the model can also achieve the effective completion of missing data and displacement trend prediction of landslide displacement time series without the need of complete original landslide displacement data. The reliability and accuracy of the model data completion and prediction are verified, further improving the feasibility of the model, which likewise helps to issue timely and accurate early warning forecast signals to remind people who are in the danger area to evacuate and avoid casualties and property damage. The method can be extended and applied in data completion and in the prediction of such landslides.

Author Contributions: Conceptualization, C.W. and Y.Z.; methodology, C.W.; validation, C.W.; formal analysis, C.W.; investigation, C.W.; resources, C.W.; data curation, C.W.; writing—original draft preparation, C.W.; writing—review and editing, C.W. and Y.Z.; supervision, Y.Z.; project administration, C.W. and Y.Z.; funding acquisition, C.W. All authors have read and agreed to the published version of the manuscript.

Funding: This research was funded by the Geological Survey Projects of China Geological Survey (No. DD20230442), the National Key Research and Development Program of China (No. 2019YFC150960101 and No. 2018YFC150480502) and the Young Scientific and Technological Talents Program of the Ministry of Natural Resources.

Institutional Review Board Statement: Not applicable.

Informed Consent Statement: Not applicable.

Data Availability Statement: Not applicable.

Conflicts of Interest: The authors declare no conflict of interest.

Abbreviations

Symbols	**Description**
$\mathcal{X} \in \mathrm{R}^{n_1 \times \cdots \times n_1}$	d-order tensor
$X_{(i)}$	The matrix that expanded along the i-th mode for tensor $\mathcal{X}$
$\| \mathcal{X} \|_*$	The kernel norm of the tensor $\mathcal{X}$
$\mathcal{Q}(\bullet)$	Transform the time series matrix into a third-order tensor
$\mathcal{P}_\Omega(Z)$	The orthogonal projection of matrix Z in the field Ω
Abbreviations	**Full Name**
MLATC	Mean-based Low-rank Autoregressive Tensor Completion
LRMC	Low-Rank Matrix Completion
HALRTC	High-Accuracy Low-Rank Tensor Completion
TRMF	Temporal Regularized Matrix Factorization
ADMM	Alternating Direction Method of Multipliers
VAR	Vector Auto-Regressive
RM	Random Missing
NM	Non-random Missing
MAPE	Mean Absolute Percentage Error
RMSE	Root-Mean-Squared Error

References

1. Wang, J.; Nie, G.; Xue, C. Landslide displacement prediction based on time series analysis and data assimilation with hydrological factors. *Arab. J. Geosci.* **2020**, *13*, 460. [CrossRef]
2. Liu, Z.-Q.; Guo, D.; Lacasse, S.; Li, J.-H.; Yang, B.-B.; Choi, J.-C. Algorithms for intelligent prediction of landslide displacements. *J. Zhejiang Univ. Sci. A* **2020**, *21*, 412–429. [CrossRef]
3. Zhang, Y.; Tang, J.; He, Z.; Tan, J.; Li, C. A novel displacement prediction method using gated recurrent unit model with time series analysis in the Erdaohe landslide. *Nat. Hazards* **2020**, *105*, 783–813. [CrossRef]
4. Li, H.; Xu, Q.; He, Y.; Deng, J. Prediction of landslide displacement with an ensemble-based extreme learning machine and copula models. *Landslides* **2018**, *15*, 2047–2059. [CrossRef]
5. Yang, B.; Yin, K.; Lacasse, S.; Liu, Z. Time series analysis and long short-term memory neural network to predict landslide displacement. *Landslides* **2019**, *16*, 677–694. [CrossRef]
6. Segoni, S.; Piciullo, L.; Gariano, S.L. Preface: Landslide early warning systems: Monitoring systems, rainfall thresholds, warning models, performance evaluation and risk perception. *Nat. Hazards Earth Syst. Sci.* **2018**, *18*, 3179–3186. [CrossRef]
7. Zhu, X.; Zhang, S.; Jin, Z.; Zhang, Z.; Xu, Z. Missing value estimation for mixed-attribute data sets. *IEEE Trans. Knowl. Data Eng.* **2011**, *23*, 110–121. [CrossRef]
8. Li, L.; Zhang, J.; Wang, Y.; Ran, B. Missing value imputation for traffic-related time series data based on a multi-view learning method. *IEEE Trans. Intell. Transp. Syst.* **2018**, *20*, 2933–2943. [CrossRef]
9. Kolda, T.G.; Bader, B.W. Tensor decompositions and applications. *SIAM Rev.* **2009**, *51*, 455–500. [CrossRef]
10. Jing, P.; Su, Y.; Jin, X.; Zhang, C. High-order temporal correlation model learning for time-series prediction. *IEEE Trans. Cybern.* **2018**, *49*, 2385–2397. [CrossRef]
11. Sen, R.; Yu, H.; Dhillon, I.S. Think globally, act locally: A deep neural network approach to high-dimensional time series forecasting. *Adv. Neural Inf. Process. Syst.* **2019**, *32*, 1–14. [CrossRef]
12. Chen, X.; Sun, L. Bayesian temporal factorization for multidimensional time series prediction. *IEEE Trans. Pattern Anal. Mach. Intell.* **2022**, *44*, 4659–4673. [CrossRef] [PubMed]
13. Yu, H.; Rao, N.; Dhillon, I.S. High-dimensional time series prediction with missing values. *arXiv* **2015**, arXiv:1509.08333.
14. Zhang, P.; Ren, P.; Liu, Y.; Sun, H. Autoregressive matrix factorization for imputation and forecasting of spatiotemporal structural monitoring time series. *Mech. Syst. Signal Process.* **2022**, *169*, 108718. [CrossRef]
15. Zhu, H.; Ding, X.; Liu, G. Rainfall prediction based on tensor complement. *Comput. Appl. Softw.* **2022**, *39*, 218–222+280. [CrossRef]
16. Chen, K.; Dong, H.; Chan, K.-S. Reduced rank regression via adaptive nuclear norm penalization. *Biometrika* **2013**, *100*, 901–920. [CrossRef]
17. Tan, H.; Wu, Y.; Shen, B.; Jin, P.J.; Ran, B. Short-term traffic prediction based on dynamic tensor completion. *IEEE Trans. Intell. Transp. Syst.* **2016**, *17*, 2123–2133. [CrossRef]
18. Figueiredo, M.; Ribeiro, B.; de Almeida, A. Electrical signal source separation via nonnegative tensor factorization using on site measurements in a smart home. *IEEE Trans. Instrum. Meas.* **2013**, *63*, 364–373. [CrossRef]
19. Li, D.; Yu, J.; Gao, W.; Chen, S.; Zhu, F. Financial time series prediction algorithm combining delay transformation and tensor decomposition. *Comput. Eng. Des.* **2022**, *43*, 1295–1303. [CrossRef]
20. Cai, J.; Candès, E.J.; Shen, Z. A singular value thresholding algorithm for matrix completion. *SIAM J. Optim.* **2010**, *20*, 1956–1982. [CrossRef]
21. Chen, X.; Sun, L. Low-rank autoregressive tensor completion for multivariate time series forecasting. *arXiv* **2020**, arXiv:2006.10436.
22. Liu, J.; Musialski, P.; Wonka, P.; Ye, J. Tensor completion for estimating missing values in visual data. *IEEE Trans. Pattern Anal. Mach. Intell* **2013**, *35*, 208–220. [CrossRef] [PubMed]
23. Zhao, H.; Shou, P.; Ma, L. A Tensor Completion Method of Missing Data in Transformer District. *Proc. CSEE* **2020**, *40*, 7328–7336. [CrossRef]
24. Candes, E.; Recht, B. Exact matrix completion via convex optimization. *Commun. ACM* **2012**, *55*, 111–119. [CrossRef]
25. Ouyang, W.; Peng, Y.; Yao, Y.; Zhang, J.; Deng, B. Anderson Acceleration for Nonconvex ADMM Based on Douglas-Rachford Splitting. *Comput. Graph. Forum* **2020**, *39*, 221–239. [CrossRef]
26. Xia, H.; Dong, Q.; Chen, Y.; Zheng, J.; Gao, C.; Wang, Z. QoS Prediction Based on the Low-Rank Autoregressive Tensor Completion. In Proceedings of the 2022 International Conference on Networking and Network Applications (NaNA), Urumqi, China, 3–5 December 2022; pp. 265–269.

applied sciences

MDPI

Article

Dual-Encoder Transformer for Short-Term Photovoltaic Power Prediction Using Satellite Remote-Sensing Data

Haizhou Cao [1,2], Jing Yang [3], Xuemeng Zhao [3], Tiechui Yao [1,2], Jue Wang [1,2,*], Hui He [3] and Yangang Wang [1,2]

1 Computer Network Information Center, Chinese Academy of Sciences, Beijing 100083, China
2 University of Chinese Academy of Sciences, Beijing 100049, China
3 School of Control and Computer Engineering, North China Electric Power University, Beijing 102206, China
* Correspondence: wangjue@sccas.cn

Abstract: The penetration of photovoltaic (PV) energy has gained a significant increase in recent years because of its sustainable and clean characteristics. However, the uncertainty of PV power affected by variable weather poses challenges to an accurate short-term prediction, which is crucial for reliable power system operation. Existing methods focus on coupling satellite images with ground measurements to extract features using deep neural networks. However, a flexible predictive framework capable of handling these two data structures is still not well developed. The spatial and temporal features are merely concatenated and passed to the following layer of a neural network, which is incapable of utilizing the correlation between them. Therefore, we propose a novel dual-encoder transformer (DualET) for short-term PV power prediction. The dual encoders contain wavelet transform and series decomposition blocks to extract informative features from image and sequence data, respectively. Moreover, we propose a cross-domain attention module to learn the correlation between the temporal features and cloud information and modify the attention modules with the spare form and Fourier transform to improve their performance. The experiments on real-world datasets, including PV station data and satellite images, show that our model achieves better results than other models for short-term PV power prediction.

Keywords: transformer; photovoltaic power forecasting; satellite images; deep learning

Citation: Cao, H.; Yang, J.; Zhao, X.; Yao, T.; Wang, J.; He, H.; Wang, Y. Dual-Encoder Transformer for Short-Term Photovoltaic Power Prediction Using Satellite Remote-Sensing Data. *Appl. Sci.* **2023**, *13*, 1908. https://doi.org/10.3390/app13031908

Academic Editor: Sergio Toscani

Received: 10 December 2022
Revised: 19 January 2023
Accepted: 31 January 2023
Published: 1 February 2023

1. Introduction

Facing the depletion of finite fossil fuels and the demands of carbon emission reduction, photovoltaic (PV) energy has raised widespread interest in recent years [1,2]. As a sustainable and flexible distributed energy source, PV power has been a significant part of the power grid with the rapid growth of its installed capacity worldwide [3]. However, the high intermittency and uncertainty of PV power make accurate prediction difficult and further pose technical challenges for reliable power system operation and control [4,5]. Meteorological factors, e.g., solar irradiance, temperature, wind speed and direction, cloud changes, and air pressure, are the primary causes of this arbitrary fluctuation. Therefore, effective PV prediction places critical demands on the model's capacity to take into account and handle various meteorological factors.

For PV power prediction, the prediction model and the forecast horizon mainly depend on the data sources. Traditional methods such as ARIMA [6] and exponential smoothing [7] hardly deal with multivariate data and thus usually take historical PV power series as the sole input. However, they are still limited to modeling non-stationary changes. To leverage the meteorological information, many previous studies mainly concentrate on employing local measurements [8] or numerical weather prediction (NWP) data [9]. However, these methods are incapable of capturing the volatility of PV power affected by cloud changes due to ignoring the data sources containing cloud conditions. More specifically, the solar irradiance reaching the PV panels mainly depends on the amount and motion of clouds [10,11] recorded in consecutive cloud images. In light of this, ground-based or satellite remote-sensing images are valuable data sources to support accurate short-term PV power prediction.

Several approaches employ ground-based/sky images for the prediction of irradiance [12,13] and PV power [14]. With sky images, the cloud motion vector (CMV) methods that can calculate the speed and direction of clouds have been developed to improve the irradiance prediction accuracy [15,16]. These methods have a considerably decreasing value beyond a 30 min future horizon and are thus commonly used in ultra-short-term PV power prediction [17]. Moreover, ground-based images are limited in observation range and require equipment installation. In contrast, satellite images offer top-down detection with greater spatial coverage, as well as advantages for prediction up to a few hours ahead [18]. They are not only suitable for CMV methods [11,19–21] but also able to provide other valuable information. Cai et al. [22] cluster the interval gray value of satellite images to construct a relationship with PV power. Wang et al. [23] propose a CNN-based fluctuation pattern prediction model with satellite images as input and apply LSTMs to individually forecast different patterns of PV power. Besides these methods that adopt only satellite images, researchers are increasingly interested in coupling satellite data with ground measurements to improve forecasting performance [24]. Si et al. [25] extracted cloud cover factor from satellite images by the modified CNN and then combined it with meteorological information for irradiance prediction. Agoua et al. [26] presented a spatiotemporal model with Lasso to integrate multi-source data for 6-h-ahead prediction. Furthermore, they used the proposed model to assess the impact of each source on forecasting performance, which showed that satellite images improved the accuracy by 3%. Yao et al. [27] proposed an encoder–decoder model, which includes U-Net and LSTM to extract the spatial features from satellite short-wave radiation data and temporal features from meteorological sequences, respectively. Then, the spatial and temporal features were concatenated for intra-hour PV power prediction.

However, a flexible predictive framework capable of handling data sources with different structures (i.e., image and sequence) is still not well discussed. Due to the insufficient consideration of the characteristics of these two structures, the spatiotemporal features extracted using the above methods are still not informative for PV power prediction up to hours. Although various deep neural networks have been applied in PV power prediction, the majority of existing methods consist of conventional deep learning models such as CNNs, RNNs, and fully connected layers. At present, few studies have followed the state-of-the-art transformer [28] architecture, which has achieved great success in natural language processing [29] and computer vision [30]. Compared with convolutional or recurrent neural networks, transformer-based models with the self-attention mechanism are superior in capturing temporal dependencies. Simeunovic et al. [31] combined a graph neural network with a transformer to propose a graph–convolutional transformer for multi-site PV power prediction based on historical weather data. In addition, most existing models based on hybrid neural networks merely concatenate the spatial and temporal features before passing them as a whole input to the following layer. As a result, this rough treatment is incapable of learning and utilizing the correlation between the features of structured local measurements and unstructured satellite images well.

Motivated by the aforementioned issues, we renovate the transformer architecture and propose a novel transformer-based PV power prediction model that can process the locally measured sequences and satellite remote-sensing images. The main contributions are summarized as follows:

1. We propose a novel **dual**-**e**ncoder **t**ransformer (DualET) for short-term PV power prediction. Distinct from the standard transformer architecture, DualET contains dual encoders, including a local seasonal information (LSI) encoder and a remote-sensing information (RSI) encoder, and a single decoder. The dual encoders are designed to handle sequence and image data, while the decoder is to model the joint meteorological features from the encoders and outputs the short-term prediction.

2. To extract informative features from the satellite images and local sequences, we deploy the two-dimensional wavelet transform block and series decomposition block in the encoding stage. The former in the RSI encoder is capable of the frequency feature extraction of image signals to obtain cloud detailed information, while the latter in the LSI encoder conducts series decomposition to improve the learning capacity of temporal patterns.
3. We propose a cross-domain attention module to learn the correlation between the temporal features in sequence data and the cloud information in image data. Furthermore, we enhance the ability to capture dependencies by modifying the attention modules with the sparse form and Fourier transform.
4. Real-world datasets are applied to evaluate the proposed model. The experiments show that our model achieves state-of-the-art results compared with other models including recent transformers.

2. Problem Definition

Two types of data sources are used in this study to predict the PV power, including cloud images processed from the satellite remote sensing data and in situ data from real PV stations. The in situ data contain historical records of several meteorological factors and PV power. The detailed data description is provided in Section 4.1. We denote the historical steps of input data as L_{in} and the prediction steps as L_{out}. The number of in situ data attributes and remote-sensing data channels used is denoted as D_{LS} and C_{RS}, respectively. In summary, this PV power forecasting problem can be formulated as

$$f(\mathbf{X}_{\text{RS}}; \mathbf{X}_{\text{LS}}) \longrightarrow \hat{\mathbf{y}} \tag{1}$$

where $f(\cdot)$ is the mapping function, i.e., the forecasting model; $\mathbf{X}_{\text{RS}} \in \mathbb{R}^{L_{\text{in}} \times H \times W \times C_{\text{RS}}}, \mathbf{X}_{\text{LS}} \in \mathbb{R}^{L_{\text{in}} \times D_{\text{LS}}}$ are the input remote-sensing data and local sequences (in situ data), respectively; $\hat{\mathbf{y}} \in \mathbb{R}^{L_{\text{out}}}$ is the PV power prediction. The hidden dimension of the model is denoted as D.

3. Methodology

PV power generation substantially fluctuates with meteorological factors, such as solar irradiance, temperature, wind speed/direction, and cloud changes. For an accurate PV power prediction, we propose a novel dual-encoder transformer (DualET) to capture context features from these factors. As shown in Figure 1, DualET has an encoder–decoder architecture like most transformers but with dual encoders. One encoder, i.e., local seasonal information encoder (LSI encoder), is to model temporal dynamics from in situ measurements. The other, i.e., remote-sensing information encoder (RSI encoder), is to learn the spatial and temporal features of clouds from satellite images. The output of the LSI encoder and that of the RSI encoder contain fluctuations in the local measurements and clouds, respectively. Both of them can be combined to provide comprehensive fluctuation information for modeling the changes in the PV power series. Further, we design a joint feature decoder to predict future short-term PV power. The details of DualET will be introduced in the following subsections.

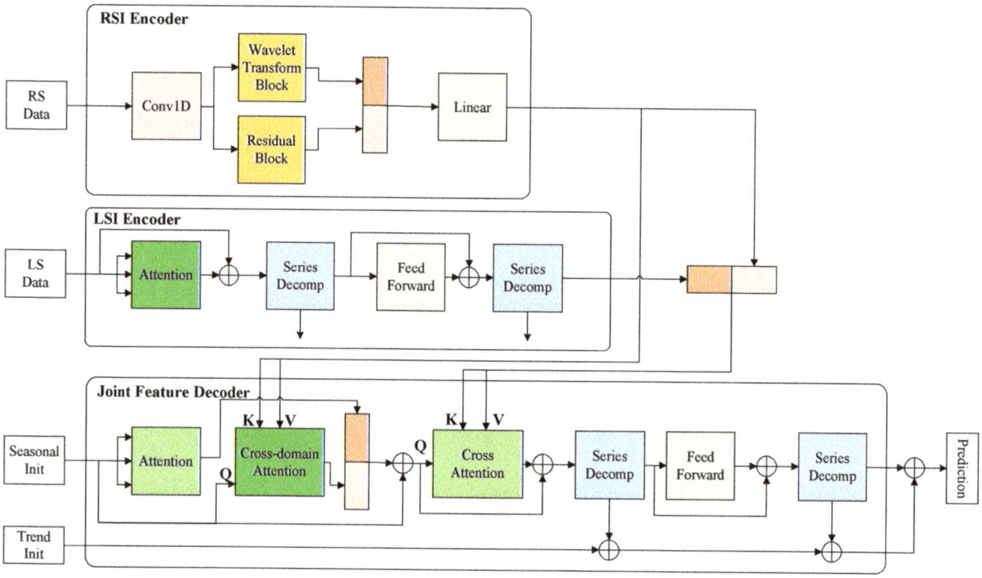

Figure 1. Model architecture of DualET.

3.1. Decomposition Modules

3.1.1. Wavelet Transform Block

The object edges represent abrupt changes around smooth regions and are distributed in the high-frequency signals of an image. For the motion and coverage of clouds, cloud edges offer crucial information. Wavelet transform is a powerful analysis tool widely used in image signal processing [32]. Compared with Fourier transform, it can capture frequency properties without location information loss. Therefore, we build the wavelet transform block (WTBlock) to extract frequency features and provide the RSI encoder with additional cloud details. As shown in Figure 2, the image signals are passed through high-pass and low-pass filters, sequentially from the horizontal and vertical directions. The high-pass filter (HPF) is to extract high-frequency components such as edges, while the low-pass filter (LPF) is to obtain low-frequency components for approximation. We summarize this 2D discrete wavelet transform as

$$\mathbf{I}_{LL}, \mathbf{I}_{LH}, \mathbf{I}_{HL}, \mathbf{I}_{HH} = \text{WTBlock}(\mathbf{I}), \quad (2)$$

where image signal $\mathbf{I}$ is decomposed into four components: the approximation $\mathbf{I}_{LL}$ (passing through LPFs in both directions), and three details ($\mathbf{I}_{LH}, \mathbf{I}_{HL}, \mathbf{I}_{HH}$) in horizontal, vertical, and diagonal orientations, respectively.

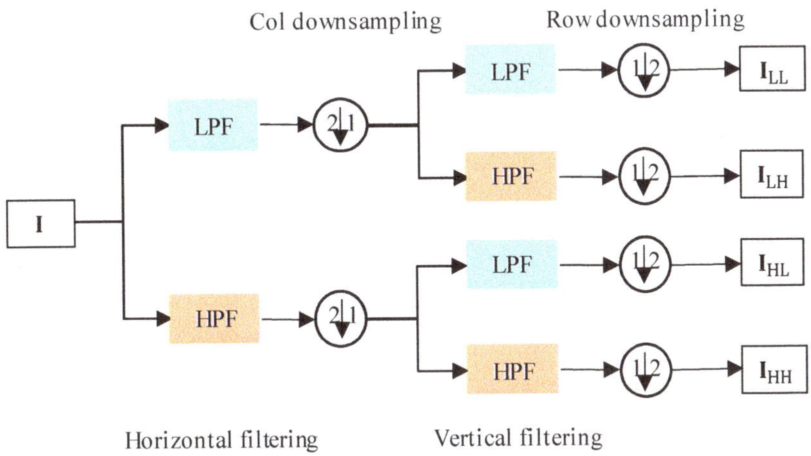

Figure 2. Illustration of 2D wavelet transform. LPF denotes the low-pass filter and HPF denotes the high-pass filter.

3.1.2. Series Decomposition Block

Real-world sequential signals (e.g., PV power sequence) are often entangled with temporal patterns that are informative for forecasting. Time series decomposition is an effective strategy to decouple knotted patterns from sequences. Among decomposition methods, seasonal-trend decomposition [33] has been widely employed as a feature engineering technique, which separates sequences into seasonal and trend parts. Inspired by Autoformer [34], we apply this decomposition idea as a series decomposition block (SDBlock) to enhance the pattern extraction ability of DualET. Given an input sequence $\mathbf{X}$, the procedure is

$$\mathbf{S}, \mathbf{T} = \text{SDBlock}(\mathbf{X}), \tag{3}$$

where $\mathbf{T}$ is the moving average result of $\mathbf{X}$ and is considered as the trend part; $\mathbf{S}$ is the residual part (i.e., the detrend part), which is regarded as the seasonal part. To keep the sequence length unchanged, a padding operation is performed on the input sequence.

3.2. Learning Modules

3.2.1. Residual Connection and Residual Block

The residual network architecture (i.e., residual connection) has become the foundation of deep neural networks, which learn residual mapping to ease the optimization of deep layers [35]. It can be formalized as $y = F(x) + x$, i.e., the input is added to the output of stacked layers (F) as the result. For the RSI encoder, we employ residual blocks to learn the representation of remote-sensing data. As shown in Figure 3, the residual block is stacked using two convolutions, batch normalization [36], and activation (ReLU) layers. The process is summarized as

$$\mathbf{X}_{\text{R}} = \text{ResBlock}(\mathbf{X}). \tag{4}$$

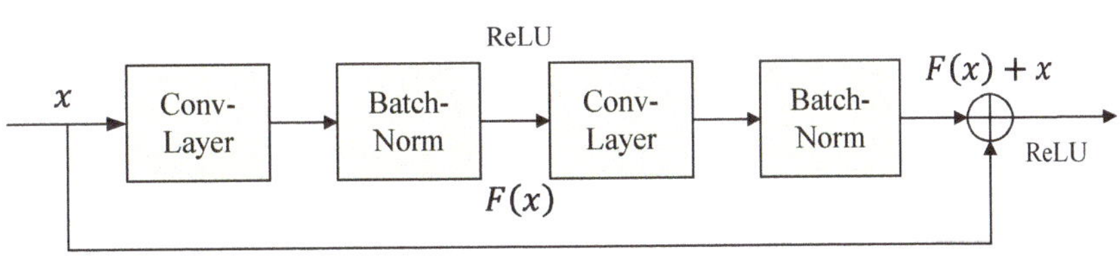

Figure 3. Illustration of residual block.

3.2.2. Attention Mechanism

As one of the most representative identifiers of transformers, the attention mechanism is proposed as a query–key–value (QKV) model to learn long-range dependencies without recurrent structures. Given the matrices $\mathbf{Q} \in \mathbb{R}^{L_q \times D_k}$, $\mathbf{K} \in \mathbb{R}^{L_k \times D_k}$, and $\mathbf{V} \in \mathbb{R}^{L_k \times D_v}$ as the projected queries, keys, and values, the single-head version of standard attention mechanism can be formalized as $\mathcal{A}(\mathbf{Q}, \mathbf{K}, \mathbf{V}) = \text{Softmax}(\frac{\mathbf{QK}^\top}{\sqrt{D_k}})\mathbf{V}$, where L_q, L_k denote the length of queries and keys/values; D_k, D_v denote the projected dimensions; $\frac{1}{\sqrt{D_k}}$ is the scale factor to avoid Softmax$(\cdot)$ yielding extremely small gradients. Furthermore, the multi-head version is as follows:

$$\begin{aligned} &\mathcal{A}_{\text{multi-head}}(\mathbf{Q}, \mathbf{K}, \mathbf{V}) = \text{Concat}(\text{head}_1, \cdots, \text{head}_H)\mathbf{W}^O, \\ &\text{where head}_i = \mathcal{A}(\mathbf{QW}_i^Q, \mathbf{KW}_i^K, \mathbf{VW}_i^V). \end{aligned} \tag{5}$$

The queries, keys, and values with dimension D are mapped into H heads (i.e., subspaces) by $\mathbf{W}_i^Q, \mathbf{W}_i^K \in \mathbb{R}^{D \times D_k}$, and $\mathbf{VW}_i^V \in \mathbb{R}^{D \times D_v}$. Then the outputs of these heads are concatenated and mapped back to D by $\mathbf{W}^O \in \mathbb{R}^{HD_v \times D}$. In most cases, $D_k = D_v = D/H$. The standard transformer has two types of multi-head attention: self-attention and cross-attention. For self-attention, its projected queries, keys, and values are shared with the same source, while the key–value pairs of cross-attention are typically from the output of encoder.

In practice, the attention modules used in DualET are modified to improve the performance of capturing dependencies. First, we design an additional cross-domain attention module to discover the correlation between the features of images and sequences. Concretely, the queries are the temporal features of the decoder, and the key–value pairs are the cloud information from the RSI encoder. Furthermore, we perform the fast Fourier transform (FFT) on the input and the inverse FFT on the output:

$$\mathcal{A}_{\text{FFT}}(\mathbf{Q}, \mathbf{K}, \mathbf{V}) = \mathcal{F}^{-1}(\mathcal{A}(\mathcal{F}(\mathbf{Q}), \mathcal{F}(\mathbf{K}), \mathcal{F}(\mathbf{V}))) \tag{6}$$

where $\mathcal{F}, \mathcal{F}^{-1}$ denote the FFT and its inverse, which are also used in the self-attention module of the LSI encoder. The FFT plays a key role in signal processing because it can rapidly convert a signal from the time/space domain to the frequency domain (and vice versa) and describe relationships between these domains [37]. It is defined by

$$\overline{\mathbf{X}}_k = \sum_{m=0}^{L-1} e^{-2\pi i k \cdot (m/L)} \mathbf{X}_m, \quad k = 0, \ldots, L-1 \tag{7}$$

Based on FFT, the attention module can discover the frequency correlation between queries and keys. In addition, we employ the ProbSparse attention mechanism [38] as the self-attention and cross-attention modules of the decoder to improve their performances.

3.2.3. Embedding and Feed-Forward Layer

For sequence modeling, the order information of time steps is crucial. Furthermore, the timestamp records of local sequences (meteorological and PV power series) are instructive for PV power prediction but are hardly utilized in the standard transformer architecture. To introduce this information, we employ timestamp-embedding layers (following Autoformer [34]) for the local sequence inputs.

The feed-forward layer is a position-wise fully connected module, i.e., the learnable parameters are shared with each step. It contains two linear layers ($\mathbf{W}_1, \mathbf{b}_1, \mathbf{W}_2, \mathbf{b}_2$) and a ReLU activation function in between, formulated as

$$\text{FeedFoward}(\mathbf{X}) = \text{ReLU}(\mathbf{XW}_1 + \mathbf{b}_1)\mathbf{W}_2 + \mathbf{b}_2. \tag{8}$$

3.3. Remote-Sensing Information Encoder

The RSI encoder is designed to learn spatial and temporal features from remote-sensing data. As shown in the top diagram of Figure 1, it is mainly composed of a two-dimensional convolution layer, a wavelet transform block, and a residual block. Given the historical L_{in} steps of cloud images $\mathbf{X}_{\text{RS}} \in \mathbb{R}^{L_{\text{in}} \times H \times W \times C_{\text{RS}}}$ as the input of the RSI encoder, the procedure is

$$\begin{aligned}
&\mathbf{I}_1 = \text{Conv2D}(\mathbf{X}_{\text{RS}}),\\
&\mathbf{I}_{\text{LL}}, \mathbf{I}_{\text{LH}}, \mathbf{I}_{\text{HL}}, \mathbf{I}_{\text{HH}} = \text{WTBlock}(\mathbf{I}_1),\\
&\mathbf{I}_2 = \text{Conv2D}(\text{ResBlock}(\mathbf{I}_1)),\\
&\mathbf{I}_3 = \text{ConvFusion}(\text{Concat}([\mathbf{I}_2, \mathbf{I}_{\text{LL}}, \mathbf{I}_{\text{LH}} + \mathbf{I}_{\text{HL}} + \mathbf{I}_{\text{HH}}])),\\
&\mathbf{Z}_{\text{RS}} = \text{Linear}(\text{Flatten}(\text{ReLU}(\mathbf{I}_3))),
\end{aligned} \tag{9}$$

where ConvFusion is 1×1 2D convolution to integrate frequency components with image features; $\mathbf{Z}_{\text{RS}} \in \mathbb{R}^{L_{\text{in}} \times D}$ is the output of the RSI encoder.

3.4. Local Seasonal Information Encoder

For weather and PV power series, the seasonal part contains the main volatility feature, which is the key to accurate prediction. Therefore, we introduce series decomposition blocks to the LSI encoder to extract seasonal patterns from local measurement data. As shown in the middle of Figure 1, the LSI encoder is staked with L_{LSI} LSI encoder layers. Given the input of the LSI encoder $\mathbf{X}^0_{\text{LS}} \in \mathbb{R}^{L_{\text{in}} \times D}$ that is embedded from $\mathbf{X}_{\text{LS}} \in \mathbb{R}^{L_{\text{in}} \times D_{\text{DS}}}$, the procedure in l-th LSI encoder layer is

$$\begin{aligned}
&\mathbf{S}^l_{\text{LS},1}, _ = \text{SDBlock}(\text{Attention}(\mathbf{X}^{l-1}_{\text{LS}}) + \mathbf{X}^{l-1}_{\text{LS}}),\\
&\mathbf{S}^l_{\text{LS},2}, _ = \text{SDBlock}(\text{FeedFoward}(\mathbf{S}^l_{\text{LS},1}) + \mathbf{S}^l_{\text{LS},1}),\\
&\mathbf{X}^l_{\text{LS}} = \mathbf{S}^l_{\text{LS},2},
\end{aligned} \tag{10}$$

where "_" is the ignored trend part; $\mathbf{S}^l_{\text{LS,i}}, i \in \{1, 2\}$ denotes the seasonal part in the l-th layer; $\mathbf{Z}_{\text{LS}} = \mathbf{X}^{L_{\text{LSI}}}_{\text{LS}}$ denotes the output of the LSI encoder.

3.5. Joint-Feature Decoder

The joint-feature decoder is to model temporal dynamics based on the joint features of local and remote-sensing data and then outputs short-term PV power. As shown in the bottom diagram of Figure 1, the decoder is stacked with L_{de} decoder layers, and each layer contains three attention modules (i.e., self-attention, cross-domain attention, and cross-attention) to determine the correlations from different perspectives. The outputs of two encoders are integrated as the input of cross-attention by ConvFusion: $\mathbf{Z}_{\text{en}} = \text{ConvFusion}(\text{Concat}([\mathbf{Z}_{\text{RS}}, \mathbf{Z}_{\text{LS}}]))$. The inputs of the decoder contain initialized seasonal part $\mathbf{X}^0_{\text{de}}$ and trend part $\mathbf{T}^0_{\text{de}}$. which are decomposed from the latter half of $\mathbf{X}^0_{\text{LS}}$ and concatenated with scalar placeholders (zeros for the seasonal part and means for the trend part). The details in the l-th decoder layer are

$$\begin{aligned}
&\mathbf{Z}^l_{\text{de},1} = \text{Attention}(\mathbf{X}^{l-1}_{\text{de}}),\\
&\mathbf{Z}^l_{\text{de},2} = \text{CrossDomianAttention}(\mathbf{X}^{l-1}_{\text{de}}, \mathbf{Z}_{\text{RS}}),\\
&\mathbf{Z}^l_{\text{de},3} = \mathbf{X}^{l-1}_{\text{de}} + \text{ConvFusion}(\text{Concat}([\mathbf{Z}^l_{\text{de},1}, \mathbf{Z}^l_{\text{de},2}])),\\
&\mathbf{S}^l_{\text{de},1}, \mathbf{T}^l_{\text{de},1} = \text{SDBlock}(\text{Attention}(\mathbf{Z}^l_{\text{de},3}, \mathbf{Z}_{\text{en}}) + \mathbf{Z}^l_{\text{de},3}),\\
&\mathbf{X}^l_{\text{de}}, \mathbf{T}^l_{\text{de},2} = \text{SDBlock}(\text{FeedFoward}(\mathbf{S}^l_{\text{de},1}) + \mathbf{S}^l_{\text{de},1}),\\
&\mathbf{T}^l_{\text{de}} = \mathbf{T}^{l-1}_{\text{de}} + \mathcal{W}^l_1 * \mathbf{T}^l_{\text{de},1} + \mathcal{W}^l_2 * \mathbf{T}^l_{\text{de},2},
\end{aligned} \tag{11}$$

where $\mathbf{Z}^l_{\text{de,i}}, i \in \{1, 2, 3\}$ is the intermediate feature; $\mathbf{S}^l_{\text{de},1}, \mathbf{X}^l_{\text{de}}$ denote the seasonal parts in l-th layer; $\mathbf{T}^l_{\text{de},1}(i \in \{1, 2\}), \mathbf{T}^l_{\text{de}}$ denote the trend parts in l-th layer; $\mathcal{W}^l_i \in \{1, 2\})$ denote the project functions for the trend parts. After L_{de} decoder layers, the final prediction $\hat{\mathbf{y}}$ is from the sum of two parts: $\mathcal{W} * \mathbf{X}_{\mathrm{de}}^{L_{\mathrm{de}}} + \mathbf{T}_{\mathrm{de}}^{L_{\mathrm{de}}}$, where $\mathcal{W}$ is the projector for the seasonal part.

4. Experiment

In this section, we evaluate the proposed DualET on satellite images and actual PV station data. We first introduce the datasets and data preprocessing. Then, we describe the experimental setting in detail. Finally, we compare the prediction performance of DualET and the baseline models and conduct several ablation experiments.

4.1. Datasets and Data Preprocessing

Two datasets were used in this study, namely, satellite remote-sensing data and PV station data. The satellite data were the L1 grid data from Himawa-8, a geostationary satellite launched in 2015 by the Japan Meteorological Agency to provide weather forecasts and typhoon and storm reports for Japan, East Asia, and the Western Pacific. The detection range of Himawa-8 is 60° N, 160° E, and 80° W, with a spatial resolution of 0.05°, which corresponds to about 5 km on the ground, and the temporal resolution is 10 min. The PV station data contains local measurements from three real PV stations at different latitudes and longitudes in Hebei, China. Each station records meteorological factors (including global and diffuse irradiance, temperature, wind direction and speed, and air pressure) and PV power records at 15 min intervals.

We set the temporal resolution as 30 min to harmonize the time intervals of these two datasets. The satellite remote-sensing data were processed into 40×40 cloud images centered on the latitude and longitude of the PV station. The satellite data were sampled from July 2018 to June 2019 to align with that of PV station data. For each day, the data with a time range from 7:00 to 19:00 (UTC +0800) were used. We divided these two datasets into a training set, a validation set, and a test set in the ratio of 8:1:1 after arranging them to ensure the test period covers multiple seasons. Before input to the model, the data were normalized using standardization to eliminate the inconsistency of the magnitude of each dimension.

4.2. Experimental Setting

4.2.1. Baseline Models

We primarily selected seven models as baselines for comparison, namely, a classic statistical model ARIMA [39], an RNN-based model LSTM [40], and three state-of-the-art models, i.e., Transformer [28] and its variants Informer [38] and Autoformer [34].

4.2.2. Hyperparameters and Platform

Our model DualET and transformer baselines, i.e., Transformer, Informer, and Autoformer, were set to the same number of layers, including two encoder layers and one decoder layer. The hidden dimension of model D was set to 512. The number of attention heads was set to 8. The batch size was set to 32, and the training epochs were set to 10 (with early stopping). The loss function was the mean-squared error (MSE) (Equation (12)), and the optimizer was ADAM with an initial learning rate of 1×10^{-4}. The input length of the model was set to 24, i.e., 12 h, and the output prediction length was set to 12, i.e., 6 h. All the models were implemented with PyTorch and conducted on a Ubuntu server with four NVIDIA GeForce RTX 2080Ti 11GB GPUs.

$$\mathrm{MSE} = \frac{1}{N}\sum_{t=1}^{N}(y_t - \hat{y}_t)^2. \tag{12}$$

4.2.3. Evaluation Metrics

We evaluated the performance of the model with three widely used metrics, i.e., mean absolute error (MAE), root-mean-squared error (RMSE), and symmetric mean absolute percentage error (SMAPE).

$$\begin{aligned} \text{MAE} &= \frac{1}{N}\sum_{t=1}^{N}|y_t - \hat{y}_t|, \\ \text{RMSE} &= \sqrt{\frac{1}{N}\sum_{t=1}^{N}(y_t - \hat{y}_t)^2}, \\ \text{SMAPE} &= \frac{100\%}{N}\sum_{t=1}^{N}\frac{|y_t - \hat{y}_t|}{(|y_t| + |\hat{y}_t|)/2}. \end{aligned} \tag{13}$$

4.3. *Results*

As shown in Table 1, we evaluated the prediction performance of the proposed DualET on three different PV stations. For short-term (6-h) PV power prediction, DualET achieved the best results on three error metrics: MAE, RMSE, and SMAPE. Compared with the results of other models on all stations, DualET achieved a relative MAE and RMSE reduction of 22.53% and 22.75%, respectively, which is a significant improvement. The average reduction in MAE was even more than 53%, compared with the traditional ARIMA, and 27.72% compared with the popular LSTM. Among these baselines, the transformer-based models, i.e., Transformer, Informer, and Autoformer, were better than ARIMA and LSTM models. Moreover, our DualET still outperformed the competitive transformer-based models and yielded a relative MAE reduction of 10.64%.

Table 1. Prediction performance of the proposed DualET.

Models	Metrics	Station0	Station1	Station2
ARIMA	MAE	3.6934	3.7727	3.6343
	RMSE	5.3633	6.3809	5.4976
	SMAPE	0.8881	0.9019	0.9135
LSTM	MAE	2.3500	2.2612	2.6287
	RMSE	3.0866	2.9739	3.4570
	SMAPE	0.7218	0.6888	0.7635
Transformer	MAE	2.0780	2.0678	1.8413
	RMSE	2.8468	2.7659	2.5875
	SMAPE	0.7110	0.6482	0.6274
Informer	MAE	1.8274	1.8393	1.8564
	RMSE	2.5874	2.5773	2.5940
	SMAPE	0.6549	0.6232	0.6498
Autoformer	MAE	1.9989	2.0690	1.9771
	RMSE	2.7379	2.7834	2.7700
	SMAPE	0.6905	0.6764	0.6810
DualET	MAE	**1.6904**	**1.7576**	**1.7657**
	RMSE	**2.2845**	**2.4205**	**2.5087**
	SMAPE	**0.6335**	**0.5931**	**0.6258**

The prediction results of different models are presented in Figures 4 and 5. It obviously shows that the number of deviation points predicted by the statistical model ARIMA is much larger than those predicted by other DNN-based models, which indicates that ARIMA is unsuitable for short-term prediction up to several hours, especially for the nonstationary PV power series. As shown in Figure 5, the LSTM model has a more scattered distribution of points than the transformer-based models, which indicates the significance of transformer architecture for sequence modeling. It can also be seen that the proposed DualET presents the fittest curves in Figure 4 and the narrowest band in Figure 5 compared with other baselines, which shows the advantages of PV power prediction.

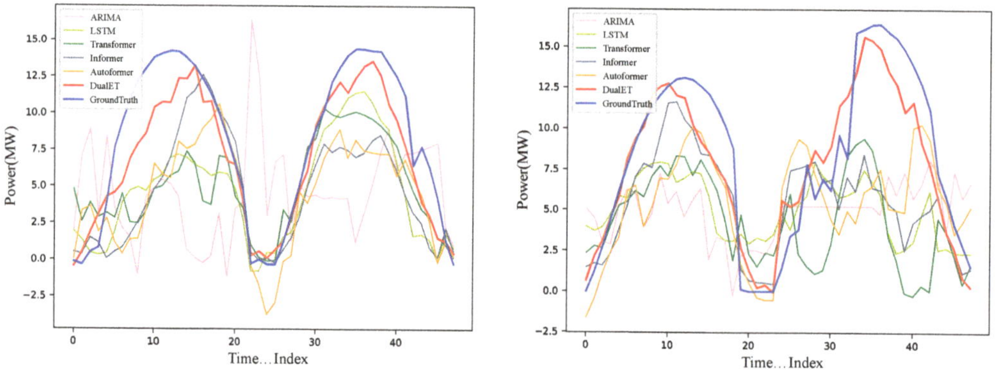

Figure 4. Prediction results of different models.

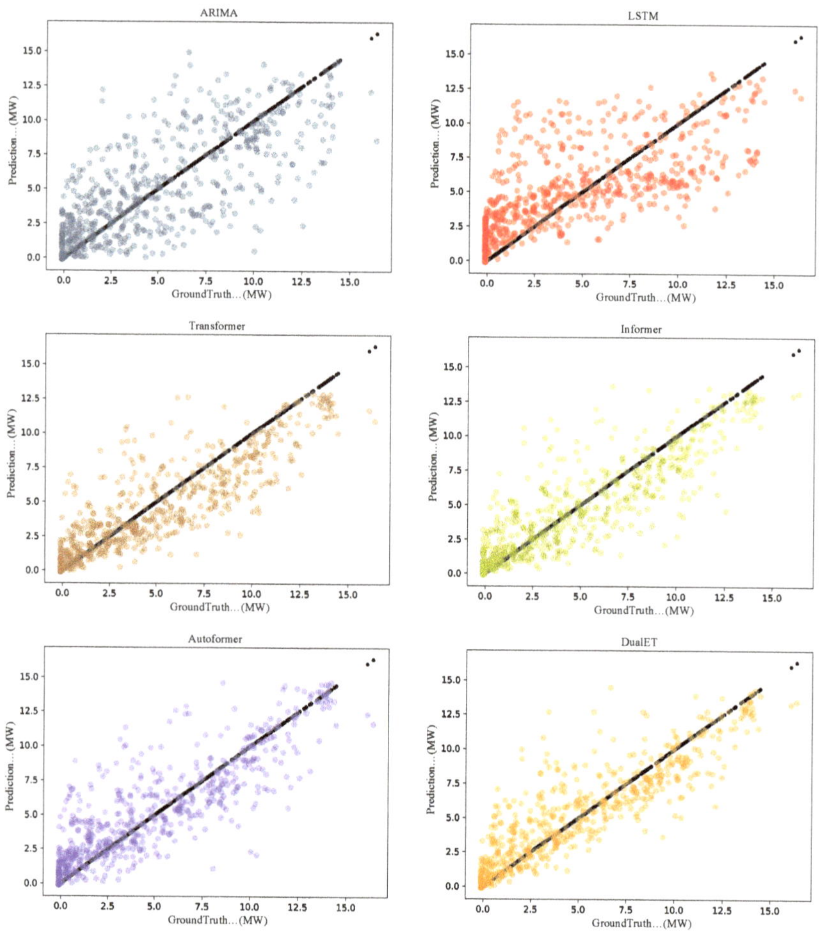

Figure 5. Scatter plots of different models.

4.4. Ablation Studies

DualET contains dual encoders, including the LSI encoder to deal with local seasonal information and the RSI encoder to process remote-sensing information, with a shared decoder to combine joint features. In addition, there are different decomposition modules and attention modules employed in DualET to enhance feature extraction. We conducted additional experiments to evaluate the impact of the dual encoders and different functional modules with MAE, RMSE, and SMAPE as evaluation metrics, and we present the results of these experiments in this section.

4.4.1. Different Encoders

We evaluated the impact of different encoders by keeping only one of the encoders from DualET. As shown in Table 2, the proposed model with only the LSI encoder is better than that with only the RSI encoder. As previously mentioned, the LSI encoder is used to learn historical seasonal information from local measurements, while the RSI encoder is to learn cloud information from satellite images. Therefore, this indicates that the fluctuation features of PV power are more contained in local measurements compared with cloud images. Meanwhile, we can see that the model with dual encoders yields the best results compared with that with only one, which shows that cloud images can provide valuable information for performance improvement and further shows the effectiveness of our dual-encoder design.

Table 2. Comparison of different encoders.

Models	MAE	RMSE	SMAPE
Only LSI encoder	1.9948	2.6964	0.6784
Only RSI encoder	2.0275	2.7453	0.6937
DualET	**1.6904**	**2.2845**	**0.6335**

4.4.2. Decomposition Blocks in Dual Encoders

We evaluated the impact of decomposition blocks for dual encoders by removing the series decomposition (SD) blocks or wavelet transform (WT) blocks. As shown in Table 3, the dual encoders without decomposition blocks will give some performance degradation. This shows that the WT and SD can enhance the learning capacity of dual encoders and provide informative features for the decoder. In addition, the performance of the RSI encoder without WT is worse than the LSI encoder without SD, which indicates that the high- and low-frequency features of cloud images are essential for short-term PV power prediction.

Table 3. Ablation studies of decomposition structures. SD and WT mean the series decomposition block and wavelet transform block, respectively. W/o means "without".

Models	MAE	RMSE	SMAPE
LSI w/o SD	1.9378	2.5927	0.6935
RSI w/o WT	1.9916	2.6440	0.7094
DualET	**1.6904**	**2.2845**	**0.6335**

4.4.3. Different Attention Modules

In DualET, the decoder's self- and cross-attention modules are formed with ProbSparse, while the others are enhanced by fast Fourier transform (FFT). As shown in Table 4, we evaluated the impact of different attention modules including the cross-domain attention module. We can see that if all attention modules are the same type (ProbSparse form or FFT), it results in performance degradation. Therefore, we employed the various attention modules with suitable modifications for better correlation capture. Moreover, the prediction error of the model without cross-domain attention significantly increases. This demonstrates the effectiveness of cross-domain attention to learn the correlation between features in sequence data and image data.

Table 4. Different attention modules. W/o means "without".

Models	MAE	RMSE	SMAPE
W/o cross-domain attention	2.0192	2.7100	1.0466
All attention with FFT	1.8060	2.4965	0.6630
All attention with ProbSparse	1.7848	2.4507	0.6657
DualET	**1.6904**	**2.2845**	**0.6335**

5. Conclusions

To handle satellite images and ground measurements, in this paper, we proposed a novel transformer-based model with dual encoders, named DualET, for short-term PV power prediction. To obtain cloud detailed information from satellite images, a two-dimensional wavelet transform block and a residual block were used in the remote-sensing information encoder. For the local seasonal information encoder, we conducted self-attention and series decomposition to learn the temporal patterns from local sequences. For the decoder, we employed three types of attention modules and series decomposition blocks to model the joint features of local and remote-sensing information and output the prediction. Specifically, a cross-domain attention module was proposed to learn the correlation between the temporal features and cloud information. Finally, the experiments on real-world datasets, including PV station data and satellite images, were presented to show the prediction performance of DualET. In addition, the ablation studies show the effectiveness of our design. In the future, we will attempt to improve the model architecture so that more data sources (e.g., NWP or other satellite remote sensing data) can be utilized to predict a longer horizon.

Author Contributions: Conceptualization, J.W. and H.H.; methodology, H.C. and J.Y.; software, H.C. and J.Y.; validation, X.Z., T.Y. and Y.W.; writing—original draft preparation, H.C. and X.Z.; writing—review and editing, T.Y., J.W. and Y.W.; visualization, J.Y. and X.Z.; supervision, J.W. and H.H. All authors have read and agreed to the published version of the manuscript.

Funding: This work was supported by the National Key R&D Program of China (2021ZD0110403).

Institutional Review Board Statement: Not applicable.

Informed Consent Statement: Not applicable.

Data Availability Statement: Not applicable.

Acknowledgments: The remote-sensing data used in this paper were processed from Himawari-8 satellite data supplied by the P-Tree System, Japan Aerospace Exploration Agency (JAXA). We also gratefully acknowledge the support of the MindSpore team.

Conflicts of Interest: The authors declare no conflict of interest. The funders had no role in the design of the study; in the collection, analyses, or interpretation of data; in the writing of the manuscript; or in the decision to publish the results.

References

1. Kabir, E.; Kumar, P.; Kumar, S.; Adelodun, A.A.; Kim, K.H. Solar Energy: Potential and Future Prospects. *Renew. Sustain. Energy Rev.* **2018**, *82*, 894–900. [CrossRef]
2. Armeanu, D.S.; Joldes, C.C.; Gherghina, S.C.; Andrei, J.V. Understanding the Multidimensional Linkages among Renewable Energy, Pollution, Economic Growth and Urbanization in Contemporary Economies: Quantitative Assessments across Different Income Countries' Groups. *Renew. Sustain. Energy Rev.* **2021**, *142*, 110818. [CrossRef]
3. Carriere, T.; Vernay, C.; Pitaval, S.; Kariniotakis, G. A Novel Approach for Seamless Probabilistic Photovoltaic Power Forecasting Covering Multiple Time Frames. *IEEE Trans. Smart Grid* **2019**, *11*, 2281–2292. [CrossRef]
4. Sanjari, M.J.; Gooi, H. Probabilistic Forecast of PV Power Generation Based on Higher Order Markov Chain. *IEEE Trans. Power Syst.* **2016**, *32*, 2942–2952. [CrossRef]

5. Stein, G.; Letcher, T.M. Integration of PV Generated Electricity into National Grids. In *A Comprehensive Guide to Solar Energy Systems*; Letcher, T.M., Fthenakis, V.M., Eds.; Academic Press: Cambridge, MA, USA, 2018; pp. 321–332. [CrossRef]
6. Li, Y.; Su, Y.; Shu, L. An ARMAX Model for Forecasting the Power Output of a Grid Connected Photovoltaic System. *Renew. Energy* **2014**, *66*, 78–89. [CrossRef]
7. Prema, V.; Rao, K.U. Development of Statistical Time Series Models for Solar Power Prediction. *Renew. Energy* **2015**, *83*, 100–109. [CrossRef]
8. Liu, L.; Zhan, M.; Bai, Y. A Recursive Ensemble Model for Forecasting the Power Output of Photovoltaic Systems. *Sol. Energy* **2019**, *189*, 291–298. [CrossRef]
9. Böök, H.; Lindfors, A.V. Site-Specific Adjustment of a NWP-Based Photovoltaic Production Forecast. *Sol. Energy* **2020**, *211*, 779–788. [CrossRef]
10. Breitkreuz, H.; Schroedter-Homscheidt, M.; Holzer-Popp, T.; Dech, S. Short-Range Direct and Diffuse Irradiance Forecasts for Solar Energy Applications Based on Aerosol Chemical Transport and Numerical Weather Modeling. *J. Appl. Meteorol. Climatol.* **2009**, *48*, 1766–1779. [CrossRef]
11. Kato, T.; Manabe, Y.; Funabashi, T.; Yoshiura, K.; Kurimoto, M.; Suzuoki, Y. A Study on Several Hours Ahead Forecasting of Spatial Average Irradiance Using NWP Model and Satellite Infrared Image. In Proceedings of the 2016 IEEE International Conference on Probabilistic Methods Applied to Power Systems (PMAPS), Beijing, China, 16–20 October 2016; pp. 1–8.
12. Zhang, C.; Du, Y.; Chen, X.; Lu, D.D.C. Cloud Motion Tracking System Using Low-Cost Sky Imager for PV Power Ramp-Rate Control. In Proceedings of the 2018 IEEE International Conference on Industrial Electronics for Sustainable Energy Systems (IESES), Hamilton, New Zealand, 31 January–2 February 2018; pp. 493–498.
13. Zhao, X.; Wei, H.; Wang, H.; Zhu, T.; Zhang, K. 3D-CNN-Based Feature Extraction of Ground-Based Cloud Images for Direct Normal Irradiance Prediction. *Sol. Energy* **2019**, *181*, 510–518. [CrossRef]
14. Lin, F.; Zhang, Y.; Wang, J. Recent Advances in Intra-Hour Solar Forecasting: A Review of Ground-Based Sky Image Methods. *Int. J. Forecast.* **2022**, *39*, 244–265. [CrossRef]
15. Bosch, J.L.; Kleissl, J. Cloud Motion Vectors from a Network of Ground Sensors in a Solar Power Plant. *Sol. Energy* **2013**, *95*, 13–20. [CrossRef]
16. Peng, Z.; Yu, D.; Huang, D.; Heiser, J.; Kalb, P. A Hybrid Approach to Estimate the Complex Motions of Clouds in Sky Images. *Sol. Energy* **2016**, *138*, 10–25. [CrossRef]
17. Tuohy, A.; Zack, J.; Haupt, S.E.; Sharp, J.; Ahlstrom, M.; Dise, S.; Grimit, E.; Mohrlen, C.; Lange, M.; Casado, M.G.; et al. Solar Forecasting: Methods, Challenges, and Performance. *IEEE Power Energy Mag.* **2015**, *13*, 50–59. . 2461351. [CrossRef]
18. Kallio-Myers, V.; Riihelä, A.; Lahtinen, P.; Lindfors, A. Global Horizontal Irradiance Forecast for Finland Based on Geostationary Weather Satellite Data. *Sol. Energy* **2020**, *198*, 68–80. [CrossRef]
19. Cros, S.; Liandrat, O.; Sebastien, N.; Schmutz, N. Extracting cloud motion vectors from satellite images for solar power forecasting. In Proceedings of the 2014 IEEE Geoscience and Remote Sensing Symposium, Quebec City, QC, Canada, 13–18 July 2014; pp. 4123–4126.
20. Kebir, N.; Maaroufi, M. Best-effort algorithm for predicting cloud motion impact on solar PV power systems production. In Proceedings of the 2018 6th International Istanbul Smart Grids and Cities Congress and Fair (ICSG), Istanbul, Turkey, 25–26 April 2018; pp. 34–38.
21. Zhou, Z.; Zhang, X.; Zhen, Z.; Mei, S. Cloud Displacement Vector Calculation in Satellite Images Based on Cloud Pixel Spatial Aggregation and Edge Matching for PV Power Forecasting. In Proceedings of the 2020 IEEE Sustainable Power and Energy Conference (iSPEC), Chengdu, China, 23–25 November 2020; pp. 112–119.
22. Cai, Y.; Liu, H.; Hu, P.; Fu, Z.; Wang, Y.; Zhang, D.; Ma, X.; Li, S. Ultra-short-term Photovoltaic Power Prediction Based on Elman Neural Network and Satellite Cloud Images. In Proceedings of the 2021 IEEE 5th Conference on Energy Internet and Energy System Integration (EI2), Taiyuan, China, 22–24 October 2021; pp. 2149–2154.
23. Wang, C.; Lu, X.; Zhen, Z.; Wang, F.; Xu, X.; Ren, H. Ultra-Short-Term Regional PV Power Forecasting Based on Fluctuation Pattern Recognition with Satellite Images. In Proceedings of the 2020 IEEE 3rd Student Conference on Electrical Machines and Systems (SCEMS), Jinan, China, 4–6 December 2020; pp. 970–975.
24. Blanc, P.; Remund, J.; Vallance, L. Short-term solar power forecasting based on satellite images. In *Renewable Energy Forecasting*; Elsevier: Amsterdam, The Netherlands, 2017; pp. 179–198.
25. Si, Z.; Yu, Y.; Yang, M.; Li, P. Hybrid Solar Forecasting Method Using Satellite Visible Images and Modified Convolutional Neural Networks. *IEEE Trans. Ind. Appl.* **2020**, *57*, 5–16. [CrossRef]
26. Agoua, X.G.; Girard, R.; Kariniotakis, G. Photovoltaic Power Forecasting: Assessment of the Impact of Multiple Sources of Spatio-Temporal Data on Forecast Accuracy. *Energies* **2021**, *14*, 1432. [CrossRef]
27. Yao, T.; Wang, J.; Wu, H.; Zhang, P.; Li, S.; Xu, K.; Liu, X.; Chi, X. Intra-Hour Photovoltaic Generation Forecasting Based on Multi-Source Data and Deep Learning Methods. *IEEE Trans. Sustain. Energy* **2021**, *13*, 607–618. [CrossRef]
28. Vaswani, A.; Shazeer, N.; Parmar, N.; Uszkoreit, J.; Jones, L.; Gomez, A.N.; Kaiser, Ł.; Polosukhin, I. Attention Is All You Need. In Proceedings of the Advances in Neural Information Processing Systems 30 (NIPS 2017), Long Beach, CA, USA, 4–9 December 2017; Volume 30.

29. Wolf, T.; Debut, L.; Sanh, V.; Chaumond, J.; Delangue, C.; Moi, A.; Cistac, P.; Rault, T.; Louf, R.; Funtowicz, M.; et al. Transformers: State-of-the-Art Natural Language Processing. In Proceedings of the 2020 Conference on Empirical Methods in Natural Language Processing: System Demonstrations, Online, 16–20 November 2020; pp. 38–45.
30. Han, K.; Wang, Y.; Chen, H.; Chen, X.; Guo, J.; Liu, Z.; Tang, Y.; Xiao, A.; Xu, C.; Xu, Y.; et al. A Survey on Vision Transformer. *IEEE Trans. Pattern Anal. Mach. Intell.* **2022**, *45*, 87–110. [CrossRef]
31. Simeunovic, J.; Schubnel, B.; Alet, P.J.; Carrillo, R.E. Spatio-Temporal Graph Neural Networks for Multi-Site PV Power Forecasting. *IEEE Trans. Sustain. Energy* **2021**, *13*, 1210–1220. [CrossRef]
32. Walker, J.S. Wavelet-Based Image Processing. *Appl. Anal.* **2006**, *85*, 439–458. [CrossRef]
33. Robert, C.; William, C.; Irma, T. STL: A Seasonal-Trend Decomposition Procedure Based on Loess. *J. Off. Stat.* **1990**, *6*, 3–73.
34. Wu, H.; Xu, J.; Wang, J.; Long, M. Autoformer: Decomposition Transformers with Auto-Correlation for Long-Term Series Forecasting. In Proceedings of the Advances in Neural Information Processing Systems, Virtual, 6–14 December 2021; Curran Associates, Inc.: Red Hook, NY, USA, 2021; Volume 34, pp. 22419–22430.
35. He, K.; Zhang, X.; Ren, S.; Sun, J. Deep Residual Learning for Image Recognition. In Proceedings of the IEEE Conference on Computer Vision and Pattern Recognition, Las Vegas, NV, USA, 27–30 June 2016; pp. 770–778.
36. Ioffe, S.; Szegedy, C. Batch Normalization: Accelerating Deep Network Training by Reducing Internal Covariate Shift. In Proceedings of the International Conference on Machine Learning, PMLR, Lille, France, 6–11 July 2015; pp. 448–456.
37. Brigham, E.O. *The Fast Fourier Transform and Its Applications*; Prentice-Hall, Inc.: Upper Saddle River, NJ, USA, 1988.
38. Zhou, H.; Zhang, S.; Peng, J.; Zhang, S.; Li, J.; Xiong, H.; Zhang, W. Informer: Beyond Efficient Transformer for Long Sequence Time-Series Forecasting. *Proc. AAAI Conf. Artif. Intell.* **2021**, *35*, 11106–11115. [CrossRef]
39. Chen, P.; Pedersen, T.; Bak-Jensen, B.; Chen, Z. ARIMA-Based Time Series Model of Stochastic Wind Power Generation. *IEEE Trans. Power Syst.* **2009**, *25*, 667–676. [CrossRef]
40. Hochreiter, S.; Schmidhuber, J. Long Short-Term Memory. *Neural Comput.* **1997**, *9*, 1735–1780. [CrossRef] [PubMed]

applied sciences

MDPI

Article

Disaster Precursor Identification and Early Warning of the Lishanyuan Landslide Based on Association Rule Mining

Junwei Xu [1,2,3], Dongxin Bai [2,3,*], Hongsheng He [1], Jianlan Luo [1] and Guangyin Lu [2,3]

1 Geophysical and Geochemical Survey Institute of Hunan Province, Changsha 410014, China
2 Key Laboratory of Metallogenic Prediction of Nonferrous Metals and Geological Environment Monitoring (Ministry of Education), School of Geosciences and Info-Physics, Central South University, Changsha 410083, China
3 Hunan Key Laboratory of Nonferrous Resources and Geological Hazards Exploration, Changsha 410083, China
* Correspondence: baidongxin07@csu.edu.cn

Abstract: It is the core prerequisite of landslide warning to mine short-term deformation patterns and extract disaster precursors from real-time and multi-source monitoring data. This study used the sliding window method and gray relation analysis to obtain features from multi-source, real-time monitoring data of the Lishanyuan landslide in Hunan Province, China. Then, the k-means algorithm with particle swarm optimization was used for clustering. Finally, the Apriori algorithm is used to mine strong association rules between the high-speed deformation process and rainfall features of this landslide to obtain short-term deformation patterns and precursors of the disaster. The data mining results show that the landslide has a high-speed deformation probability of more than 80% when rainfall occurs within 24 h and the cumulative rainfall is greater than 130.60 mm within 7 days. It is of great significance to extract the short-term deformation pattern of landslides by data mining technology to improve the accuracy and reliability of early warning.

Keywords: disaster precursor identification; early warning; association rule mining; particle swarm optimization; k-means clustering; Apriori algorithm; gray relation analysis

Citation: Xu, J.; Bai, D.; He, H.; Luo, J.; Lu, G. Disaster Precursor Identification and Early Warning of the Lishanyuan Landslide Based on Association Rule Mining. *Appl. Sci.* **2022**, *12*, 12836. https://doi.org/10.3390/app122412836

Academic Editors: Jinrong Jiang, Yangang Wang and Yuzhu Wang

Received: 9 November 2022
Accepted: 12 December 2022
Published: 14 December 2022

1. Introduction

Mountains and hills make up more than 60% of the total area of Hunan province in China, half of which have slopes greater than 25° [1]. This area has high rainfall, so landslide disasters are frequent. According to statistics, 2449 various geological disasters occurred in Hunan Province in 2020, causing economic losses of 262.49 million RMB, of which 2116 were landslide disasters, accounting for 86.4% [2]. Deploying multiple types of sensors on landslides to gather information on deformation, rainfall, stress, and other physical parameters, and providing timely warning, are low-cost and reliable prevention methods that can effectively reduce casualties [3–5]. With the development of sensor technology and Internet of Things technology, landslide monitoring is gradually developing towards the direction of automation and intelligence [6–9]. It is of great significance to fully mine extensive monitoring data and extract and identify warning precursors for studying the mechanisms of landslide disasters and improving the accuracy of warning.

Early and accurate identification of landslide precursors is a prerequisite for early warning. The traditional precursors that can be used for early warning are mainly macroscopic phenomena such as surface cracks, slope toe uplift and other macro phenomena [10–12]. With the development of monitoring technology, landslide precursors can be mined from abundant monitoring data, of which the most widely used type of data is surface deformation. The accelerated deformation process of landslides is the most intuitive and reliable precursor, so it is widely used in the study of landslide early warning. Xu et al. [5,13] proposed to use the normalized tangent angle as an indicator for early warning of landslides.

Jeng et al. [14] proposed to use displacement-velocity ratio as an indicator for landslide warning. Valletta et al. [15] proposed a multicriteria approach to identify accelerated deformation processes in landslides. Bai et al. [16] proposed a hybrid warning algorithm that could identify the landslide acceleration process quickly, automatically, and accurately in an online monitoring and warning system, and achieved the balance of warning immediacy, accuracy, and computational resources through different strategies.

Although displacement, as a precursor of landslide disaster, can give early warning quickly and accurately, it also has many shortcomings. First, the current sensors for displacement monitoring are highly susceptible to environmental influences and often generate false alarms during the warning process [16–18]. Second, displacement is the result of a combination of multiple factors, both internal and external to the landslide. The acceleration of displacement foreshadows the initiation of the landslide process, and the warning window is very short [19–21]. Finally, the use of a single displacement characteristic for early warning does not take into account the impact of external trigger factors such as rainfall, earthquakes, and construction on the disaster, and is therefore necessarily incomplete.

The development of data mining technology in recent years has provided new research ideas for landslide precursor identification. Data mining technology can filter and analyze useful information and important events from massive data to reveal the internal relationships and hidden rules of data, which have been widely used in the commercial [22,23], industrial [24,25], engineering [26,27], medical [28–30] and educational [31,32] fields with remarkable effect. The application of data mining techniques in the field of landslides is mainly focused on susceptibility assessment [33–35], aiming to analyze landslide instability risk at the regional scale, while there are very few studies on application in specific landslide monitoring. Ma et al. [36,37] first used modern data mining techniques integrating two-step clustering, association rule mining, and decision trees to analyze data from the Majiagou landslide and the Zhujiadian landslide in the Three Gorges reservoir area. These studies not only identified landslide disaster factors but also realized the prediction of displacement evolution, which was the earliest research to carry out data mining for single landslide monitoring. Miao et al. [38] and Guo et al. [39] adopted the same data mining technology to analyze the trigger factors of the Baishuihe landslide and the Shuping landslide in the Three Gorges Reservoir area, and determined the warning threshold. All these studies have fully and comprehensively considered the correlation between multi-source monitoring data and provided causal relationships between different monitoring variables, which are very helpful for the analysis of landslide damage mechanisms and instability patterns. Most of these studies focused on reservoir landslides in the Three Gorges region of China, with monitoring data collected over several years and on a monthly scale. Therefore, these studies were more focused on the long-term deformation patterns of landslides. However, the daily-scale or even hourly-scale short-term deformation patterns of landslides are equally important in landslide early warning studies. Such short-term deformation patterns contain more reliable precursors of landslide disasters than deformation features, which are important for early warning decisions. In addition, these studies all adopted a two-step clustering algorithm, which is a kind of hierarchical clustering and divides clusters through the process of splitting or clustering, so there is no need to determine the number of clusters. However, for the clustering of daily or even hourly monitoring data, we prefer to flexibly adjust the number of clusters. This kind of data is very complex, and human subjective judgment is still needed. At this time, partition clustering represented by k-means is more appropriate.

The purpose of this paper was to mine the short-term deformation patterns of landslides, identify the precursors of landslides, and obtain more reliable early warnings. In this study, the Lishanyuan Landslide in Hunan Province was taken as the case study. First, the sliding window method was used to extract features from the original monitoring data, then the k-means algorithm optimized by particle swarm optimization (PSO) was used to cluster the features and construct the item set, and the Apriori algorithm was finally used to mine the association rules between different features and determine the short-term deformation pattern of landslides according to the given confidence levels to analyze the precursors of landslide disasters and provide early warnings.

2. Methodology

2.1. Overview

The association mining method as shown in Figure 1 was used to mine the association rules between the triggering factors and landslide displacement, which mainly includes three parts: feature engineering, clustering and association rule mining.

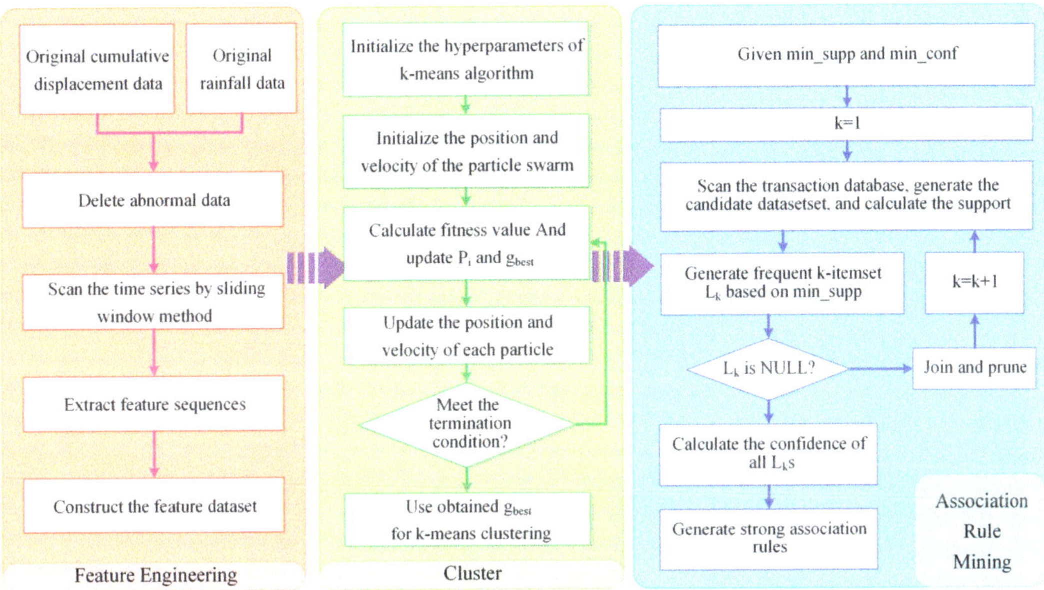

Figure 1. Flowchart of association rule mining for the Lishanyuan landslide.

In the feature engineering part, for the original multi-source data obtained from landslide monitoring, the sliding window method is used to scan the monitoring data time series of each source. In the scanning process, the 3σ criterion is first used to eliminate obvious outliers, and then the corresponding features are calculated according to the type of monitoring data, and finally the feature time series data set is formed.

In the clustering part, for the feature time series obtained in the previous part, the PSO-optimized k-means algorithm is first used for clustering, and then the time series are transformed into item sets, and finally the time series of all features are processed in the same way to build the transaction database.

In the association rule mining part, for the transaction database constructed in the previous section, the Apriori algorithm is used to mine the frequent item sets and association rules in the transaction database and analyze the disaster factors and destabilization precursors of landslides accordingly.

2.2. PSO-Optimized k-Means Algorithm

The original value-based monitoring dataset must be changed into a category-based transactional database since the Apriori algorithm for association rule mining is category-based. The k-means algorithm is the most well-known clustering algorithm, whose core objective is to classify the dataset into K clusters, with the elements in each cluster having a high degree of similarity. The k-means algorithm is simple to implement and fast to cluster, but it is very sensitive to the choice of initial cluster centers. Different initial values may lead to different clustering results, i.e., local optima rather than global optima. To solve this problem, we used the PSO algorithm for global optimization. The PSO algorithm is an evolutionary algorithm based on population intelligence that finds the optimal solution by simulating the process of a flock of birds searching for food. The specific steps of the k-means clustering algorithm optimized by PSO are as follows:

Step 1: Particle swarm initialization. Suppose there is a particle swarm composed of m particles in a given D-dimensional search space, and each particle has only two attributes: position and velocity, where position is the code of the solution to be solved and the velocity is the iteration step size.

For the $i-th$ particle, its coordinate position can be expressed as:

$$X_i = (x_{i1}, \ x_{i2}, \ \cdots, \ x_{iD}) \tag{1}$$

The velocity of the $i-th$ particle can be expressed as:

$$V_i = (v_{i1}, \ v_{i2}, \ \cdots, \ v_{iD}) \tag{2}$$

When performing k-means clustering on the dataset $D = \{x_1, \ x_2, \ \cdots, \ x_n\}$, the initial cluster centers $C = \{\mu_1, \ \mu_2, \ \cdots, \ \mu_k\}$ need to be specified. In order to avoid the problem of local optimal clustering caused by the sensitivity of C, we coded C as X_i in Equation (1) for global optimization.

Step 2: Particle clustering and fitness calculation. Perform k-means clustering after decoding each particle in the particle swarm. The specific steps are as follows:

Sub-step 2.1: For each element x_i in the dataset D, the Euclidean distance $d_{ij} = \sqrt{\sum_{i=1}^{n} (x_i - \mu_j)^2}$ between x_i and the center μ_j of each cluster is calculated and the current element x_i is assigned to the cluster C_j represented by the center with the smallest distance.

Sub-step 2.2: For each cluster C_j obtained in Sub-step 2.1, the central $\mu_j' = \frac{1}{|C_j|} \sum_{x \in C_j} x$ of that cluster is recalculated and the $C = \{\mu_1, \ \mu_2, \ \cdots, \ \mu_k\}$ is updated.

Sub-step 2.3: Repeat the sub-step 2.1 and 2.2 until the center μ_j' and element x_i of each cluster C_j no longer change. Then, the final clustering result can be obtained.

Sub-step 2.4: To evaluate the clustering effect of the current position of each particle, the following equation is used to calculate the fitness $F(i)$ of each particle.

$$F(i) = \sum_{i=1}^{n} \sum_{j=1}^{k} (x_i - \mu_j)^2 \tag{3}$$

where x_i denotes the $i-th$ element in the dataset, and μ_j is the center of the $i-th$ cluster. The fitness function represents the sum of the squares of the distances between each element and the center of the cluster to which the element belongs, and the smaller the fitness, the better the clustering effect. The individual optimal solution P_i and the group optimal solution g_{best} can be obtained through fitness.

The optimal position searched by the $i-th$ particle is denoted as:

$$P_i = (p_{i1}, \ p_{i2}, \ \cdots, \ p_{iD}) \tag{4}$$

The optimal position searched by the particle swarm is denoted as:

$$g_{best} = (g_1, \ g_2, \ \cdots, \ g_D) \tag{5}$$

Step 3: Position update. Update the position and velocity of each particle with the following equation:

$$V_i^{k+1} = \omega V_i^k + c_1 r_1 \left(P_i^k - X_i^k\right) + c_2 r_2 \left(g_{best}^k - X_i^k\right) \quad (6)$$

$$X_i^{k+1} = X_i^k + V_i^{k+1} \quad (7)$$

where V_i^k denotes the velocity of the $i-th$ particle at the $k-th$ iteration. X_i^k denotes the position of the $i-th$ particle at the $k-th$ iteration. P_i^k denotes the individual optimal solution of the $i-th$ particle up to the $k-th$ iteration. g_{best}^k denotes the population optimal solution of the particle swarm as of the $k-th$ iteration. c_1 and c_2 denote the acceleration constants to adjust the step size. r_1 and r_2 denote the random numbers between 0 and 1, respectively, to enhance the randomness of the search process.

Step 4: After the velocity and position of each particle are updated, the particles that are out of the solution range are initialized randomly again. If the current fitness function value is better than the historical optimal P_i, then update P_i. Similarly, if the population fitness function value of the updated particle population is better than the historical optimal g_{best}, then update g_{best}.

Step 5: Repeat Step 2 to 4, and constantly update and iterate for all particles until the maximum number of solutions is reached or the aggregation degree σ^2 of the group optimal solution g_{best} is less than the given threshold.

$$\sigma^2 = \frac{1}{n}\sum_{i=1}^{n}\left[F(i) - \overline{F}\right]^2 \quad (8)$$

where $\overline{F}$ is the average fitness of the particle swarm, and σ^2 represents the aggregation degree of the particles in the particle swarm. The smaller its value, the higher the convergence degree of the PSO algorithm. When σ^2 is less than the given threshold, this means that the particles are all clustered near the global solution. At this time, the particle with the best fitness is the initial center of the global optimal clustering, and the clustering result can be obtained using k-means for clustering.

2.3. Association Rule Mining and Apriori Algorithm

Association rule mining refers to the discovery of valuable correlation information and knowledge rules from data sets. The Apriori algorithm is the most classic algorithm for mining association rules. Suppose $I = \{i_1, i_2, \cdots, i_m\}$ is an item set, each element i_m of which is called an item, and the item set of length k is called $k - itemset$. A subset of item set I can form a transaction, and multiple transactions can form a transaction database $T = \{t_1, t_2, \cdots, t_n\}$. Suppose X and Y are two item sets in the transaction database whose intersection is empty, that is, $X \subset T, Y \subset T$ and $X \cap Y = \varnothing$. These two item sets can be denoted by $X \Rightarrow Y$ and if there is an association rule the former item X denotes the condition of the association rule and the latter item Y denotes the conclusion of the association rule. To better measure the performance of the mined association rules, three indicators need to be used: support, confidence and lift. Their definitions are as follows:

Support is the probability that X and Y occur together in the transaction database T. Support indicates the importance of association rule $X \Rightarrow Y$ in the total data:

$$S_{X \Rightarrow Y} = \frac{|T(X \cup Y)|}{|T|} \quad (9)$$

Confidence is the probability that Y will occur if X is included. Confidence expresses the validity of the association rule $X \Rightarrow Y$:

$$C_{X \Rightarrow Y} = \frac{|T(X \cup Y)|}{|T(X)|} \quad (10)$$

Lift is the ratio of the confidence to the occurrence probability of the later term Y in the transaction database T. Lift indicates the strength of the correlation, and the larger the lift, the stronger the correlation:

$$L_{X \Rightarrow Y} = \frac{|T(X \cup Y)|}{|T(X)|} / \frac{|T(Y)|}{|T|} \tag{11}$$

where $|T(X \cup Y)|$ represents the number of item sets X and Y appearing in the transaction database T at the same time. $|T|$ represents the number of transactions in the transaction database T. $|T(X)|$ and $|T(Y)|$ represent the number of item sets X or Y appearing in the transaction database T, respectively.

The minimum support min_supp and minimum confidence min_conf need to be specified as thresholds in association rule mining. If the support of an item set is greater than min_supp, then this item set is called frequent item set. If the support and confidence of an association rule are greater than the min_supp and min_conf, then this rule is called a strong association rule. The specific flow of the Apriori algorithm is shown in Figure 1 and described in detail as follows:

Step 1: Iterate through all the transactions in the transaction database T and count the number of each item and calculate the support. The items with the support greater than min_supp are deleted to generate the frequent 1-item set L_1.

Step 2: Generate candidate 2-item set for L_1 by joining and pruning operations, calculate the support of each item in the candidate 2-item set and also filter according to the min_supp to get the frequent 2-item set L_2. Repeat this process until the candidate $k - itemset$ is empty, thus obtaining the frequent $k - itemset$.

Step 3: Calculate the confidence of each L_k separately, and output the association rules with confidence greater than min_conf.

3. Study Area

3.1. Landslide Overview

The Lishanyuan landslide is located in Xinhua County, Hunan Province, China (Figure 2). The longitudinal length of the landslide is 120 m, the horizontal width is 300 m, the average thickness is about 3 m, and the total volume is about 1.08×10^5 m^3. The landslide is a shallow landslide with a main slide direction of 210°. The middle and back edges of the slope are well covered with vegetation. There are several residential houses at the left foot of the slope. The area on the right side of the slope is poorly covered with vegetation. There is a village-level road and a small stream at the front edge of the landslide, and the foot of the slope has been washed by the river for a long time. Due to long-term river scouring at the foot of the slope, the landslide initially showed accelerated deformation characteristics in 1996. From then until 2012, it underwent a slow deformation trend. In 2013, the landslide accelerated again, with multiple cracks on the slope and subsidence of the village-level road. In April 2018, affected by heavy rainfall, the landslide had a local slip of about 600 m^3, and the sliding soil fell to the walls and windows of residential houses on the lower side of the slope, causing a direct loss of about 600,000 RMB. According to the on-site investigation, the landslide is a small and shallow traction landslide, which is very common and representative in Hunan Province, China.

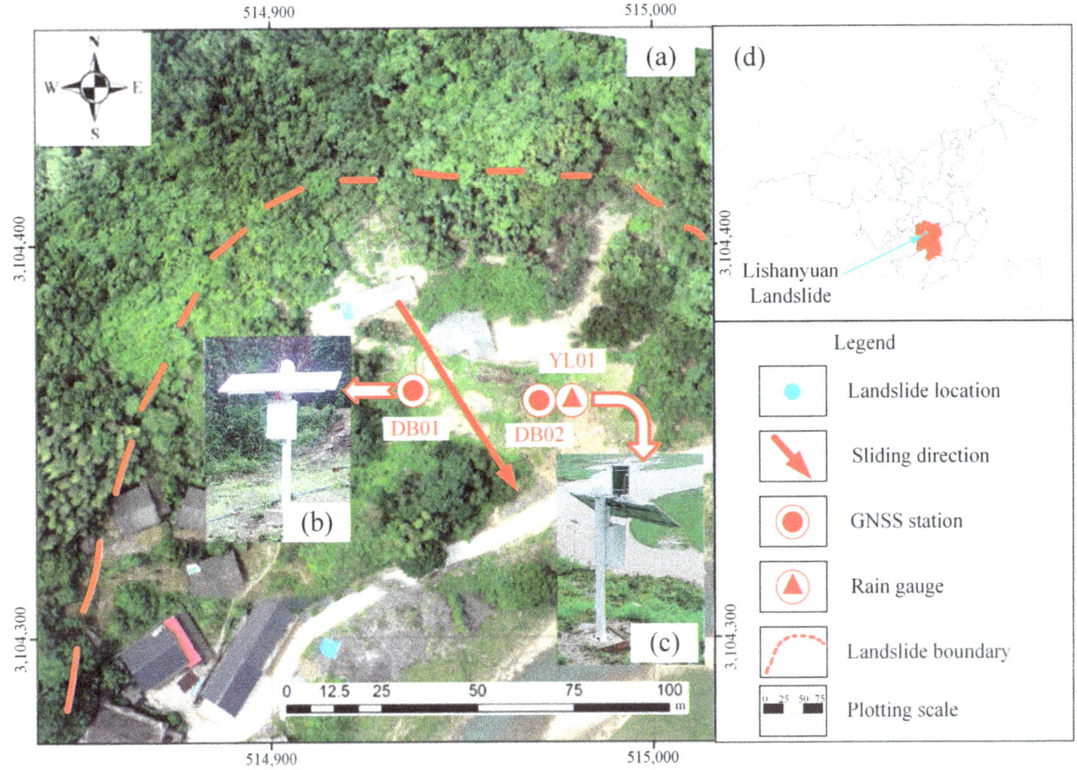

Figure 2. Geographical location and monitoring scheme of Lishanyuan landslide. (**a**) Site photograph of the Lishanyuan landslide. (**b**) Geographical location of the Lishanyuan landslide. (**c**) Photographs of monitoring stations DB02 and YL01. (**d**) Photograph of the DB01 monitoring station.

3.2. Deformation Characteristics

To protect the safety of the residents below the landslide, we completed the deployment and commissioning of monitoring equipment to establish a monitoring and early warning system on 15 April 2021. The location and photos of the monitoring stations are shown in Figure 2. Two GNSS monitoring stations, named DB01 and DB02, were deployed on the main slide profile of the landslide, and the GNSS base stations are located on the roadside of the lower side of the landslide. In addition, a rain gauge named YL01 was deployed at DB02. The automated monitoring system received the first monitoring data at 17:00 on 15 April, and the default acquisition interval of the GNSS monitoring stations was 1 h. As the landslide appeared to accelerate significantly on 17 May, the GNSS monitoring stations adjusted the collection interval to 30 min, and the collection interval of the rain gauge was adjusted to 20 min. As of 15:00 p.m. on 1 July 2022, a total of 57,597 monitoring data were collected by the monitoring system, including 30,396 GNSS monitoring data and 27,201 rainfall monitoring data. The monitoring data are shown in Figure 3.

From Figure 3, it can be seen that the deformation patterns of the two GNSS monitoring stations are basically the same, but the deformation amplitude of DB02 is significantly larger than that of DB01, which indicates that the deformation of the leading edge of this landslide is larger than that of the trailing edge of the landslide, which is consistent with the deformation characteristics of the traction landslide. The threshold design and warning process of this landslide are described in Bai et al. [16] The deployed monitoring and warning system is able to accurately and quickly identify the accelerated deformation process of the landslide and report timely warnings. To verify the reliability of the monitoring data, we inspected the landslide site on 19 May 2021. At this time, the landslide area had just experienced a strong rainfall, and the monitoring data from two GNSS monitoring stations showed that the landslide had been violently deformed. We found multiple cracks in the landslide body during a site inspection (Figure 4), obvious slippage, soil accumulation at the foot of the slope, and small mudslides in the local area. These macroscopic phenomena are consistent with the monitoring and early warning results, proving the effectiveness and reliability of the monitoring and early warning system.

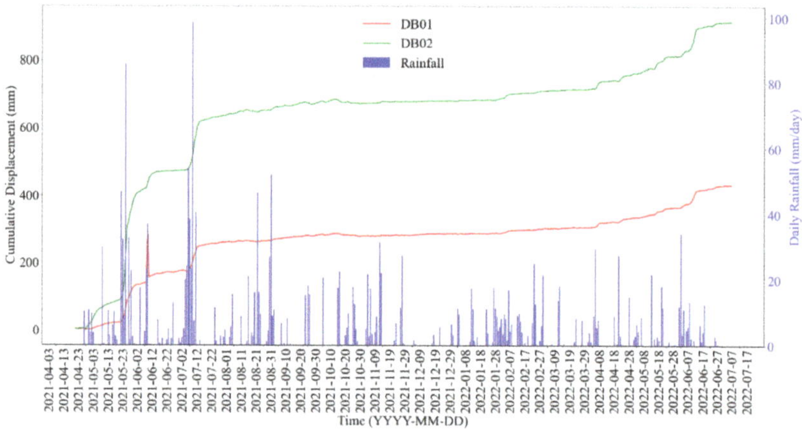

Figure 3. Daily rainfall data and displacement data from two GNSS monitoring stations for the Lishanyuan landslide.

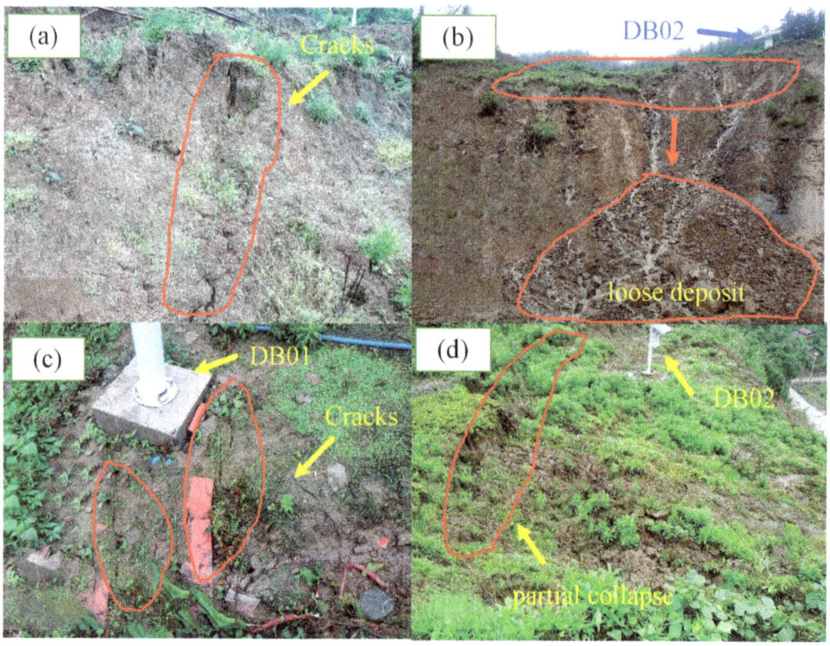

Figure 4. On-site inspection photos on 19 May 2021. (**a**) Long cracks on the surface of the landslide. (**b**) Loose deposit near the DB02 station. (**c**) Multiple cracks near DB01 station. (**d**) partial collapse near DB02 station.

The deformation process of Lishanyuan landslide shows obvious correlation with the rainfall process. Taking the DB02 monitoring station with the most obvious deformation as an example, the displacement of the two GNSS monitoring stations first showed a fluctuation of 10 mm for about a week after the monitoring started, indicating that the measurement accuracy of GNSS was of centimeter level. Affected by the rainfall event on 22 April 2021, the acceleration process began with the synchronization of the displacements of the two GNSS monitoring stations starting at 4:00 a.m. on 23 April. After that, the displacements of the two monitoring stations showed a step-like growth, and each severe deformation process was accompanied by concentrated high-intensity rainfall. After mid-October, the rainfall decreased, and the deformation began to slow down, showing creep characteristics. After April of the following year, the landslides started a process of obvious deformation and acceleration again.

3.3. Feature Engineering

From the deformation characteristics reflected by the Lishanyuan landslide monitoring data, we found that the deformation process of the landslide showed an obvious correlation with the rainfall process. To further mine the association rules of this correlation, we needed to carry out further data mining on the monitoring data, for which feature engineering was first needed. Feature engineering refers to extracting more representative features from raw monitoring data to improve the effectiveness of mining tasks. For the monitoring data and deformation characteristics of the Lishanyuan landslide, we constructed features for both deformation and velocity. In terms of deformation, we focused more on the accelerated deformation process, so the deformation velocity was the most important feature. The deformation velocity (v_{DB01},v_{DB02}) of two GNSS monitoring stations was chosen as the main feature. In terms of rainfall, we paid attention not only to the short-term rainfall features, but also to the long-term rainfall features. We chose the cumulative rainfall of three hours q^{3h}, six hours q^{6h}, twelve hours q^{12h}, 24 h q^{24h}, three days q^{3d}, and seven days q^{7d} as the characteristics reflecting rainfall.

According to Bai et al. [40] and Liu et al. [41], the strength of correlation between features can be quantitatively determined by gray relation analysis. Therefore, we used the gray relation analysis algorithm to calculate the gray relation degree between various types of rainfall features and deformation velocity; the calculation results are shown in Table 1. From Table 1, we can see that the gray relation degree of all rainfall features and deformation velocity is greater than 0.9, which is much higher than the empirical threshold of 0.6. So, all of these rainfall features can be adopted.

Table 1. Gray relation degree between rainfall characteristics and displacement characteristics.

	q^{3h}	q^{6h}	q^{12h}	q^{24h}	q^{3d}	q^{7d}
v_{DB01}	0.970735	0.971045	0.971962	0.973857	0.978478	0.979868
v_{DB02}	0.964633	0.964742	0.96582	0.968061	0.973537	0.975926

4. Results

4.1. Clustering Results

For the various types of feature sequences obtained from feature engineering, we used the PSO-optimized k-means algorithm to cluster each feature. The number of cluster centers for each type of feature was set to 3, thereby clustering the feature into low, medium, and high clusters. The clustering results of all features are shown in Figure 5, and the interval ranges and sample sizes of different clusters are shown in Table 2.

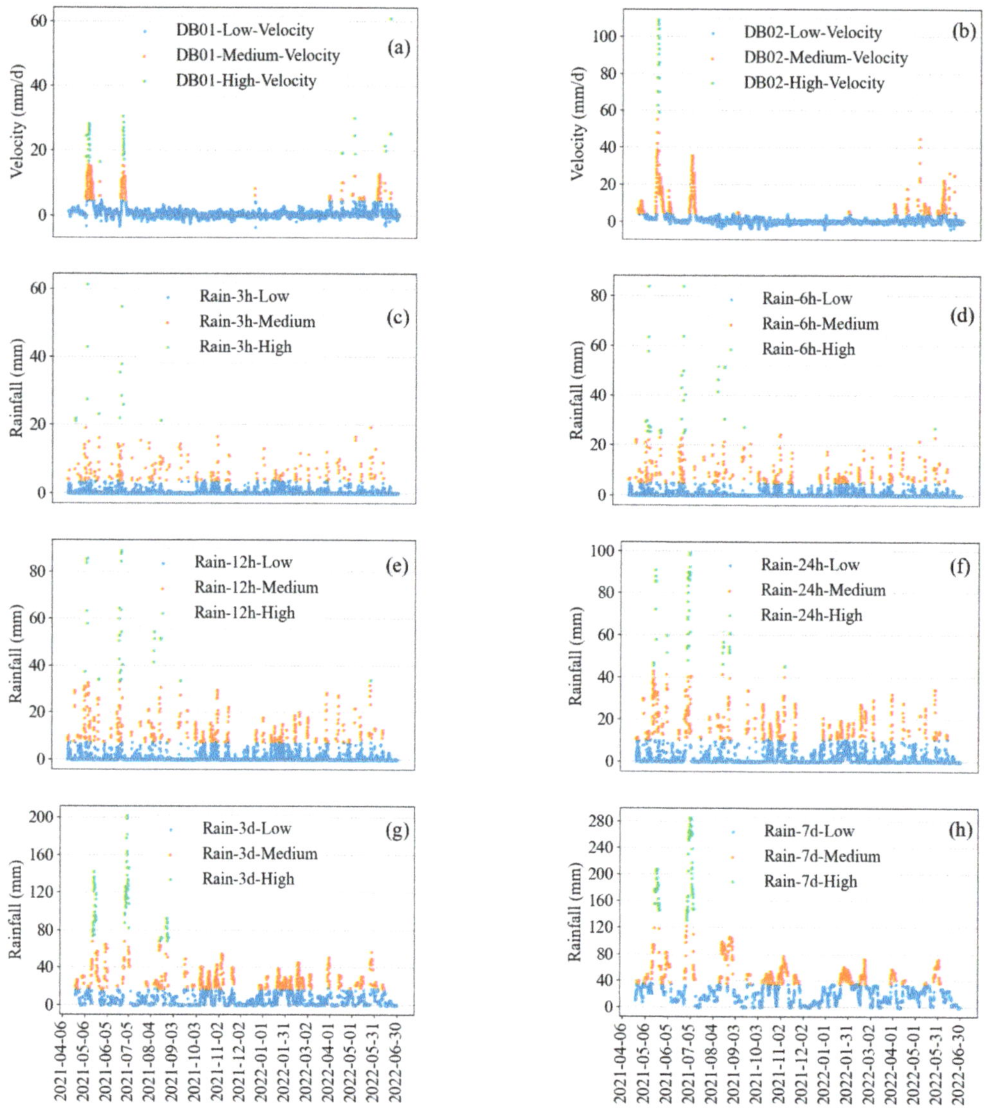

Figure 5. Visualization of all feature clustering results. (**a**) The velocity of DB01. (**b**) The velocity of DB02. (**c**) 3-h cumulative rainfall. (**d**) 6-h cumulative rainfall. (**e**) 12-h cumulative rainfall. (**f**) 24-h cumulative rainfall. (**g**) 3-day cumulative rainfall. (**h**) 7-day cumulative rainfall.

Table 2. Interval range and sample size of all feature clustering results.

Feature Name	Cluster Name	Lower Bound	Upper Bound	Count	Mean	Standard Deviation
v_{DB01}	DB01-Low-Velocity	−3.64	4.70	4887	0.46	1.02
	DB01-Medium-Velocity	4.78	15.54	253	9.09	2.83
	DB01-High-Velocity	16.49	60.96	56	23.58	6.25
v_{DB02}	DB02-Low-Velocity	−3.84	4.77	4669	0.60	1.23
	DB02-Medium-Velocity	4.79	55.37	507	13.39	8.55
	DB02-High-Velocity	59.24	108.84	20	90.67	15.85
q^{3h}	Rain-3 h-Low	0.00	3.60	4962	0.18	0.56
	Rain-3 h-Medium	3.80	19.40	217	7.33	3.43
	Rain-3 h-High	21.20	61.40	17	34.34	12.17
q^{6h}	Rain-6 h-Low	0.00	5.00	4830	0.32	0.88
	Rain-6 h-Medium	5.20	24.40	333	9.95	4.57
	Rain-6 h-High	24.80	83.80	33	39.42	16.47
q^{12h}	Rain-12 h-Low	0.00	7.20	4656	0.64	1.48
	Rain-12 h-Medium	7.40	32.80	494	13.93	6.12
	Rain-12 h-High	33.40	89.20	46	52.80	17.83
q^{24h}	Rain-24 h-Low	0.00	10.40	4429	1.39	2.55
	Rain-24 h-Medium	10.60	43.00	698	19.64	7.76
	Rain-24 h-High	45.00	99.40	69	68.48	17.00
q^{3d}	Rain-3 d-Low	0.00	17.20	3736	4.31	5.12
	Rain-3 d-Medium	17.40	68.20	1284	30.46	11.61
	Rain-3 d-High	68.60	202.20	176	106.48	28.89
q^{7d}	Rain-7 d-Low	0.00	35.20	3554	15.46	10.88
	Rain-7 d-Medium	35.40	122.20	1450	55.17	17.47
	Rain-7 d-High	130.60	285.80	192	197.53	46.03

Combining Figure 5 and Table 2, it can be seen that the number of samples in different clusters differs by an order of magnitude. The number of samples of low-rank clusters is much higher than that of middle-rank and high-rank clusters, and the number of samples of middle-rank clusters is also much higher than that of high-rank clusters. Combining Figure 5 and Table 2, it can be seen that the number of samples in different clusters differs by an order of magnitude. The number of samples of low-rank clusters is much higher than that of middle-rank and high-rank clusters, and the number of samples of middle-rank clusters is also much higher than that of high-rank clusters. Taking v_{DB01} as an example, the speed of samples in the DB01-Low-Velocity cluster is between −3.64 and 4.70, which has a total of 4887 samples. The speed of samples in the DB01-Medium-Velocity cluster is between 4.78 and 15.54 with a total of 253 samples, which is an order of magnitude less than the DB01-Low-Velocity cluster. The speed of samples in the DB01-High-Velocity cluster is between 16.49 and 60.96, and the number of samples is only 56, which is an order of magnitude less than the DB01-Medium-Velocity cluster. The clustering results of other features have similar characteristics to v_{DB01}, differing only in the range of intervals. The boundaries between the different clusters are very clear, and the characterized velocities or intensities of rainfall are largely consistent with the actual situation.

4.2. Association Rule Mining Results

After clustering, each cluster is named, and then the values in the features converted into category names. The category names of different features at each moment form an item set, thereby transforming the entire feature dataset into a transaction database. The Apriori algorithm was used to carry out the association rule mining study on this transaction database to mine strong association rules between rainfall features and the velocities of two GNSS monitoring stations separately. We took the velocity of GNSS monitoring stations as the latter term and the rainfall characteristics as the former term, and obtained the corresponding strong association rules based on both different min_conf and min_supp. For the velocity of the DB01 monitoring station, we set the min_supp as 0.3% and the min_conf as 80%. For landslide warning, we focused more on the high-speed deformation process, which is the DB01-High-Velocity cluster, so we filtered the eligible association rules as shown in Table 3.

Table 3. Association rules related to Lishanyuan landslide deformation.

Rule ID	Mined Association Rules	Confidence	Support	Lift
1	Rain-24 h-Low & Rain-3 d-High & Rain-7 d-High => DB01-High-Velocity	86.36%	0.37%	80.13
2	Rain-12 h-Low & Rain-24 h-Low & Rain-3 d-High & Rain-7 d-High => DB01-High-Velocity	86.36%	0.37%	80.13
3	Rain-24 h-Low & Rain-3 d-High & Rain-3 h-Low & Rain-7 d-High => DB01-High-Velocity	90.48%	0.37%	83.95
4	Rain-24 h-Low & Rain-3 d-High & Rain-6 h-Low & Rain-7 d-High => DB01-High-Velocity	86.36%	0.37%	80.13
5	Rain-12 h-Low & Rain-24 h-Low & Rain-3 d-High & Rain-3 h-Low & Rain-7 d-High => DB01-High-Velocity	90.48%	0.37%	83.95
6	Rain-12 h-Low & Rain-24 h-Low & Rain-3 d-High & Rain-6 h-Low & Rain-7 d-High => DB01-High-Velocity	86.36%	0.37%	80.13
7	Rain-24 h-Low & Rain-3 d-High & Rain-3 h-Low & Rain-6 h-Low & Rain-7 d-High => DB01-High-Velocity	90.48%	0.37%	83.95
8	Rain-12 h-Low & Rain-24 h-Low & Rain-3 d-High & Rain-3 h-Low & Rain-6 h-Low & Rain-7 d-High =>DB01-High-Velocity	90.48%	0.37%	83.95
9	Rain-12 h-Low & Rain-24 h-High & Rain-7 d-High => DB02-High-Velocity	83.33%	0.10%	216.50
10	Rain-12 h-Low & Rain-24 h-High & Rain-3 d-High & Rain-7 d-High => DB02-High-Velocity	83.33%	0.10%	216.50
11	Rain-12 h-Low & Rain-24 h-High & Rain-3 h-Low & Rain-7 d-High => DB02-High-Velocity	83.33%	0.10%	216.50
12	Rain-12 h-Low & Rain-24 h-High & Rain-6 h-Low & Rain-7 d-High => DB02-High-Velocity	83.33%	0.10%	216.50
13	Rain-12 h-Low & Rain-24 h-High & Rain-3 d-High & Rain-3 h-Low & Rain-7 d-High => DB02-High-Velocity	83.33%	0.10%	216.50
14	Rain-12 h-Low & Rain-24 h-High & Rain-3 d-High & Rain-6 h-Low & Rain-7 d-High => DB02-High-Velocity	83.33%	0.10%	216.50
15	Rain-12 h-Low & Rain-24 h-High & Rain-3 h-Low & Rain-6 h-Low & Rain-7 d-High => DB02-High-Velocity	83.33%	0.10%	216.50
16	Rain-12 h-Low & Rain-24 h-High & Rain-3 d-High & Rain-3 h-Low & Rain-6 h-Low & Rain-7 d-High => DB02-High-Velocity	83.33%	0.10%	216.50

For the velocity of DB02 monitoring station, we set the min_supp as 0.1% and the min_conf as 80%. We also filtered the association rules with DB01-High-Velocity as the latter term in the same way (see Table 3).

A lot of interesting information can be obtained from the association rules in Table 3. First, the lift of all these association rules is much greater than 1, indicating that the presence of rainfall former terms in these association rules has a significant positive effect on the high-speed deformation of landslides. Second, if the rainfall characteristics are classified into the current moment (3 h, 6 h), short-term (12 h, 24 h), and long-term (3 days, 7 days), then the recent rainfall characteristics are not significant in the association rules. For example, in Rules 3–8 and 11–16, these association rules with recent rainfall characteristics can be considered as subordinate rules of the four main rules: Rule 1, Rule 2, Rule 9, and Rule 10. Third, from the four main rules of Rule 1, Rule 2, Rule 9, and Rule 10, the high-speed deformation of landslides requires not only the occurrence of short-term rainfall characteristics, but also long-term rainfall characteristics, and the occurrence of only one of them does not induce the high-speed deformation process of landslides. Fourth, for the DB01 monitoring station, the long-term heavy rainfall characteristics are more important for high-speed deformation of the landslide, because the three-day or seven-day rainfall characteristics in Rule 1–8 are heavy rainfall, and the 12- and 24-h rainfall characteristics can be low-intensity rainfall. Fifth, for the DB02 monitoring station, not only the long-term heavy rainfall characteristics of 3–7 days but also the short-term heavy rainfall characteristics of 24 h are required.

In conclusion, by analyzing the monitoring data of the Lishanyuan landslide, it can be initially concluded that the landslide is caused by rainfall. Through association rule mining, the disaster factors can be more accurately identified as the combination of short-term rainfall and long-term heavy rainfall. When making early warning decisions, a rainfall within 24 h and a heavy rainfall with a cumulative rainfall greater than 130.60 mm within 7 days can be used as a precursor to identify the high-speed deformation of the landslide.

5. Discussion

To analyze the disaster factors of the Lishanyuan landslide and determine the precursors of high-speed deformation of the landslide, we used a combination of PSO-optimized k-means clustering algorithm and the Apriori algorithm to mine the association rules of the monitoring data. The analysis results of the mined strong association rules show that the high-speed deformation process of the Lishanyuan landslide is mainly affected by the combination of short-term rainfall of about 1 day, and long-term heavy rainfall of about 3–7 days. A rainfall within 24 h and a heavy rainfall with a cumulative rainfall greater than 130.60 mm within 7 days can be used as a precursor to identify the high-speed deformation of the landslide. Such a precursor can improve the ability of warning.

The association rule mining algorithm used in this study has the following main advantage. First, we used the sliding window method to extract features in the feature engineering part. This method improves the data utilization by considering continuous data over a period of time comprehensively, compared to considering only the features at the current moment, thus improving the reliability and representativeness of the obtained features. Second, the original k-means clustering algorithm is optimized by using the PSO algorithm, which effectively prevents the clustering results from falling into a local optimal. Third, the k-means algorithm is simple to implement and only requires a given number of clusters, which is easy to quantify. Other clustering methods that do not require specifying the number of clusters often require specifying other hyperparameters that are difficult to quantify. It is more convenient to directly specify the number of clusters for the control of clustering results. Finally, this study is based on real-time monitoring data, whose sampling intervals are hourly or even on the minute scale. Compared with ultra-long-term monitoring data at the monthly scale, it is richer and pays more attention to short-term deformation patterns of landslides, which is of great significance for early warning.

Additionally, it should be noted that our improvement of the association rule mining method results in an increase in algorithm complexity. On the one hand, we use the PSO algorithm to optimize the k-means clustering process, which is an evolutionary algorithm that requires uninterrupted iterative computation of many potential solutions, which is very complex and time-consuming. On the other hand, the Apriori algorithm for mining association rules needs to scan the entire transaction database when processing frequent candidate sets, which has high algorithm complexity, a huge amount of calculation, and is very time-consuming. With the improvement of technology and the passage of monitoring time, the number of monitored landslides and the volume of data will also increase sharply in the future. It is an inevitable trend to explore simple and fast data mining algorithms.

In this study, the Apriori algorithm was used to mine association rules. Therefore, the numerical dataset was converted into a category-type transaction database. This method cannot further quantify association rules and is easily affected by clustering results. Meanwhile, the Apriori algorithm does not consider the time series characteristics of item sets in the mining process of association rules, which results in ignoring the influence of sequence pattern in the mining process. Future research needs to explore a data mining method that uses numerical datasets and considers sequential patterns in order to mine more valuable information.

6. Conclusions

For the monitoring data of the Lishanyuan landslide, the sliding window method was used to extract the features, and gray relation analysis was used to screen the features. Then the PSO-optimized k-means algorithm was used to cluster. Finally, the Apriori algorithm was used to mine the strong association rules between deformation speed and rainfall characteristics to analyze the disaster factors of the Lishanyuan landslide and propose the precursors that can be used for early warning. The following conclusions were obtained from this study:

The sliding window method was adopted to achieve feature extraction of high-frequency monitoring data, which can make full use of the data and be more representative.

Using PSO-optimized k-means algorithm to cluster feature engineering can effectively avoid the clustering results falling into local optimal. By clustering, the numerical dataset is transformed into transaction database, and the strong association rules can be mined using the Apriori algorithm. This research developed mining of association rules of monitoring data at hourly or even minute scale. Compared with ultra-long-term monitoring data at monthly scale, we should pay more attention to short-term deformation patterns, which are more conducive to short-term real-time early warning.

The results of association rules mining show that the high-speed deformation process of the Lishanyuan landslide is mainly affected by the combination of short-term rainfall of about 1 day and long-term heavy rainfall of about 3–7 days. A rainfall within 24 h and heavy rainfall with a cumulative rainfall greater than 130.60 mm within 7 days can be used as a precursor to identify the high-speed deformation of the landslide.

The association rule mining algorithm used in this paper is highly complex, computationally intensive, and very time-consuming, and simpler and faster algorithms need to be explored in the future to cope with monitoring and early warning of more and more landslides. In addition, this mining process does not consider the time-series characteristics of item sets, and future research should explore sequence pattern mining, which has uncovered more and more valuable information.

Author Contributions: Conceptualization, J.X., D.B.; methodology, J.X., D.B.; software, J.X., D.B.; validation, J.X. and J.L.; formal analysis, J.X.; investigation, J.X.; resources, J.X. and H.H.; data curation, J.X.; writing—original draft preparation, J.X.; writing—review and editing, J.X., H.H., J.L., D.B., G.L.; visualization, D.B. and J.X.; supervision, J.X.; project administration, J.X.; funding acquisition, G.L. All authors have read and agreed to the published version of the manuscript.

Funding: This research was funded by Key research and development program of Hunan Province of China, grant number: 2020SK2135. Natural Resources Research Project in Hunan Province of China, grant number: 2021-15.

Institutional Review Board Statement: Not applicable.

Informed Consent Statement: Not applicable.

Data Availability Statement: Not applicable.

Acknowledgments: We would like to thank the editor and the reviewers for helping us improve the quality of the manuscript.

Conflicts of Interest: The authors declare no conflict of interest.

References

1. Yu, D.; Hu, S.; Tong, L.; Xia, C. Spatiotemporal Dynamics of Cultivated Land and Its Influences on Grain Production Potential in Hunan Province, China. *Land* **2020**, *9*, 510. [CrossRef]
2. National Bureau of Statistics of the People's Republic of China. *China Statistical Yearbook-2021*; China Statistics Press: Beijing, China, 2021.
3. Bai, D.; Tang, J.; Lu, G.; Zhu, Z.; Liu, T.; Fang, J. The Design and Application of Landslide Monitoring and Early Warning System Based on Microservice Architecture. *Geomat. Nat. Hazards Risk* **2020**, *11*, 928–948. [CrossRef]
4. Chen, M.; Jiang, Q. An Early Warning System Integrating Time-of-Failure Analysis and Alert Procedure for Slope Failures. *Eng. Geol.* **2020**, *272*, 105629. [CrossRef]
5. Xu, Q.; Peng, D.; Zhang, S.; Zhu, X.; He, C.; Qi, X.; Zhao, K.; Xiu, D.; Ju, N. Successful Implementations of a Real-Time and Intelligent Early Warning System for Loess Landslides on the Heifangtai Terrace, China. *Eng. Geol.* **2020**, *278*, 105817. [CrossRef]
6. Liu, Y.; Tang, G.; Zou, W. Video Monitoring of Landslide Based on Background Subtraction with Gaussian Mixture Model Algorithm. In Proceedings of the 2021 IEEE International Geoscience and Remote Sensing Symposium IGARSS, Brussels, Belgium, 11–16 July 2021; pp. 8432–8435.
7. Lau, Y.M.; Wang, K.L.; Wang, Y.H.; Yiu, W.H.; Ooi, G.H.; Tan, P.S.; Wu, J.; Leung, M.L.; Lui, H.L.; Chen, C.W. Monitoring of Rainfall-Induced Landslides at Songmao and Lushan, Taiwan, Using IoT and Big Data-Based Monitoring System. *Landslides* **2022**, 1–26. [CrossRef]
8. Sheikh, M.R.; Nakata, Y.; Shitano, M.; Kaneko, M. Rainfall-Induced Unstable Slope Monitoring and Early Warning through Tilt Sensors. *Soils Found.* **2021**, *61*, 1033–1053. [CrossRef]

9. Liu, C.; Shao, X.; Li, W. Multi-Sensor Observation Fusion Scheme Based on 3D Variational Assimilation for Landslide Monitoring. *Geomat. Nat. Hazards Risk* **2019**, *10*, 151–167. [CrossRef]
10. Fan, X.; Xu, Q.; Liu, J.; Subramanian, S.S.; He, C.; Zhu, X.; Zhou, L. Successful Early Warning and Emergency Response of a Disastrous Rockslide in Guizhou Province, China. *Landslides* **2019**, *16*, 2445–2457. [CrossRef]
11. Zhu, L.; Deng, Y.; He, S. Characteristics and Failure Mechanism of the 2018 Yanyuan Landslide in Sichuan, China. *Landslides* **2019**, *16*, 2433–2444. [CrossRef]
12. Ma, S.; Xu, C.; Xu, X.; He, X.; Qian, H.; Jiao, Q.; Gao, W.; Yang, H.; Cui, Y.; Zhang, P.; et al. Characteristics and Causes of the Landslide on July 23, 2019 in Shuicheng, Guizhou Province, China. *Landslides* **2020**, *17*, 1441–1452. [CrossRef]
13. Xu, Q.; Yuan, Y.; Zeng, Y.; Hack, R. Some New Pre-Warning Criteria for Creep Slope Failure. *Sci. China Technol. Sci.* **2011**, *54*, 210–220. [CrossRef]
14. Jeng, C.J.; Chen, S.S.; Tseng, C.H. A Case Study on the Slope Displacement Criterion at the Critical Accelerated Stage Triggered by Rainfall and Long-Term Creep Behavior. *Nat. Hazards* **2022**, *112*, 2277–2312. [CrossRef]
15. Valletta, A.; Carri, A.; Segalini, A. Definition and Application of a Multi-Criteria Algorithm to Identify Landslide Acceleration Phases. *Georisk Assess. Manag. Risk Eng. Syst. Geohazards* **2021**, *16*, 555–569. [CrossRef]
16. Bai, D.; Lu, G.; Zhu, Z.; Zhu, X.; Tao, C.; Fang, J. A Hybrid Early Warning Method for the Landslide Acceleration Process Based on Automated Monitoring Data. *Appl. Sci.* **2022**, *12*, 6478. [CrossRef]
17. Tan, Q.; Wang, P.; Hu, J.; Zhou, P.; Bai, M.; Hu, J. The Application of Multi-Sensor Target Tracking and Fusion Technology to the Comprehensive Early Warning Information Extraction of Landslide Multi-Point Monitoring Data. *Measurement* **2020**, *166*, 108044. [CrossRef]
18. Li, W.; Tsung, F.; Song, Z.; Zhang, K.; Xiang, D. Multi-Sensor Based Landslide Monitoring via Transfer Learning. *J. Qual. Technol.* **2021**, *53*, 474–487. [CrossRef]
19. Bai, D.; Lu, G.; Zhu, Z.; Zhu, X.; Tao, C.; Fang, J. Using Electrical Resistivity Tomography to Monitor the Evolution of Landslides' Safety Factors under Rainfall: A Feasibility Study Based on Numerical Simulation. *Remote Sens.* **2022**, *14*, 3592. [CrossRef]
20. Denchik, N.; Gautier, S.; Dupuy, M.; Batiot-Guilhe, C.; Lopez, M.; Léonardi, V.; Geeraert, M.; Henry, G.; Neyens, D.; Coudray, P.; et al. In-Situ Geophysical and Hydro-Geochemical Monitoring to Infer Landslide Dynamics (Pégairolles-de-l'Escalette Landslide, France). *Eng. Geol.* **2019**, *254*, 102–112. [CrossRef]
21. Jiang, Y.; Xu, Q.; Lu, Z.; Luo, H.; Liao, L.; Dong, X. Modelling and Predicting Landslide Displacements and Uncertainties by Multiple Machine-Learning Algorithms: Application to Baishuihe Landslide in Three Gorges Reservoir, China. *Geomat. Nat. Hazards Risk* **2021**, *12*, 741–762. [CrossRef]
22. Qing, H.; Zheng, G.; Fu, D. Risk Data Analysis of Cross Border E-Commerce Transactions Based on Data Mining. *J. Phys. Conf. Ser.* **2021**, *1744*, 032014. [CrossRef]
23. Tornero-Velez, R.; Isaacs, K.; Dionisio, K.; Prince, S.; Laws, H.; Nye, M.; Price, P.S.; Buckley, T.J. Data Mining Approaches for Assessing Chemical Coexposures Using Consumer Product Purchase Data. *Risk Anal.* **2021**, *41*, 1716–1735. [CrossRef]
24. Espadinha-Cruz, P.; Godina, R.; Rodrigues, E.M.G. A Review of Data Mining Applications in Semiconductor Manufacturing. *Processes* **2021**, *9*, 305. [CrossRef]
25. Dogan, A.; Birant, D. Machine Learning and Data Mining in Manufacturing. *Expert Syst. Appl.* **2021**, *166*, 114060. [CrossRef]
26. Liu, W.; Wang, H.; Xi, Z.; Zhang, R.; Huang, X. Physics-Driven Deep Learning Inversion with Application to Magnetotelluric. *Remote Sens.* **2022**, *14*, 3218. [CrossRef]
27. Guo, Y.; Cui, Y.; Xie, J.; Luo, Y.; Zhang, P.; Liu, H.; Liu, J. Seepage Detection in Earth-Filled Dam from Self-Potential and Electrical Resistivity Tomography. *Eng. Geol.* **2022**, *306*, 106750. [CrossRef]
28. Hua, S.; Liu, Q.; Yin, G.; Guan, X.; Jiang, N.; Zhang, Y. Research on 3D Medical Image Surface Reconstruction Based on Data Mining and Machine Learning. *Int. J. Intell. Syst.* **2022**, *37*, 4654–4669. [CrossRef]
29. Ishaq, A.; Sadiq, S.; Umer, M.; Ullah, S.; Mirjalili, S.; Rupapara, V.; Nappi, M. Improving the Prediction of Heart Failure Patients' Survival Using SMOTE and Effective Data Mining Techniques. *IEEE Access* **2021**, *9*, 39707–39716. [CrossRef]
30. Wu, W.-T.; Li, Y.-J.; Feng, A.-Z.; Li, L.; Huang, T.; Xu, A.-D.; Lyu, J. Data Mining in Clinical Big Data: The Frequently Used Databases, Steps, and Methodological Models. *Mil. Med. Res.* **2021**, *8*, 44. [CrossRef]
31. Palacios, C.A.; Reyes-Suárez, J.A.; Bearzotti, L.A.; Leiva, V.; Marchant, C. Knowledge Discovery for Higher Education Student Retention Based on Data Mining: Machine Learning Algorithms and Case Study in Chile. *Entropy* **2021**, *23*, 485. [CrossRef]
32. Xin, Y. Analyzing the Quality of Business English Teaching Using Multimedia Data Mining. *Mob. Inf. Syst.* **2021**, *2021*, e9912460. [CrossRef]
33. Yong, C.; Jinlong, D.; Fei, G.; Bin, T.; Tao, Z.; Hao, F.; Li, W.; Qinghua, Z. Review of Landslide Susceptibility Assessment Based on Knowledge Mapping. *Stoch Environ. Res Risk Assess* **2022**, *36*, 2399–2417. [CrossRef]
34. Rafiei Sardooi, E.; Azareh, A.; Mesbahzadeh, T.; Soleimani Sardoo, F.; Parteli, E.J.R.; Pradhan, B. A Hybrid Model Using Data Mining and Multi-Criteria Decision-Making Methods for Landslide Risk Mapping at Golestan Province, Iran. *Environ. Earth Sci.* **2021**, *80*, 487. [CrossRef]
35. Vakhshoori, V.; Pourghasemi, H.R.; Zare, M.; Blaschke, T. Landslide Susceptibility Mapping Using GIS-Based Data Mining Algorithms. *Water* **2019**, *11*, 2292. [CrossRef]

36. Ma, J.; Tang, H.; Liu, X.; Hu, X.; Sun, M.; Song, Y. Establishment of a Deformation Forecasting Model for a Step-like Landslide Based on Decision Tree C5.0 and Two-Step Cluster Algorithms: A Case Study in the Three Gorges Reservoir Area, China. *Landslides* **2017**, *14*, 1275–1281. [CrossRef]
37. Ma, J.; Tang, H.; Hu, X.; Bobet, A.; Zhang, M.; Zhu, T.; Song, Y.; Ez Eldin, M.A.M. Identification of Causal Factors for the Majiagou Landslide Using Modern Data Mining Methods. *Landslides* **2017**, *14*, 311–322. [CrossRef]
38. Miao, F.; Wu, Y.; Li, L.; Liao, K.; Xue, Y. Triggering Factors and Threshold Analysis of Baishuihe Landslide Based on the Data Mining Methods. *Nat. Hazards* **2021**, *105*, 2677–2696. [CrossRef]
39. Guo, L.; Miao, F.; Zhao, F.; Wu, Y. Data Mining Technology for the Identification and Threshold of Governing Factors of Landslide in the Three Gorges Reservoir Area. *Stoch. Environ. Res. Risk Assess* **2022**, *36*, 3997–4012. [CrossRef]
40. Bai, D.; Lu, G.; Zhu, Z.; Tang, J.; Fang, J.; Wen, A. Using Time Series Analysis and Dual-Stage Attention-Based Recurrent Neural Network to Predict Landslide Displacement. *Environ. Earth Sci.* **2022**, *81*, 509. [CrossRef]
41. Liu, Q.; Lu, G.; Dong, J. Prediction of Landslide Displacement with Step-like Curve Using Variational Mode Decomposition and Periodic Neural Network. *Bull. Eng. Geol. Environ.* **2021**, *80*, 3783–3799. [CrossRef]

applied sciences

MDPI

Article

Submarine Landslide Susceptibility and Spatial Distribution Using Different Unsupervised Machine Learning Models

Xing Du [1,2,*], Yongfu Sun [3], Yupeng Song [1,*], Zongxiang Xiu [1] and Zhiming Su [1]

[1] First Institute of Oceanography, MNR, Qingdao 266061, China
[2] College of Environmental Science and Engineering, Ocean University of China, Qingdao 266100, China
[3] National Deep Sea Center, Qingdao 266237, China
* Correspondence: duxing@fio.org.cn (X.D.); songyupeng@fio.org.cn (Y.S.)

Abstract: A submarine landslide is a well-known geohazard that can cause significant damage to offshore engineering facilities. Most standard predicting and mapping methods require expert knowledge, supervision, and fieldwork. In this research, the main objective was to analyze the potential of unsupervised machine learning methods and compare the performance of three different unsupervised machine learning models (k-means, spectral clustering, and hierarchical clustering) in modeling the susceptibility of the submarine landslide. Nine groups of geological factors were selected as the input parameters, which were obtained through field surveys. To estimate submarine landslide susceptibility, all input factors were separated into three or four groups based on data features and environmental variables. Finally, the goodness-of-fit and accuracy of models were validated with both internal metrics (Calinski–Harabasz index, silhouette index, and Davies–Bouldin index) and external metrics (existing landslide distribution, hydrodynamic distribution, and liquefication distribution). The findings of k-means, spectral clustering, and hierarchical clustering performed commendably and accurately in forecasting the submarine landslide susceptibility. Spectral clustering has the greatest congruence with environmental geology parameters. Therefore, the unsupervised machine learning model can be used in submarine-landslide-predicting studies, and the spectral clustering method performed best. Furthermore, machine learning can improve submarine landslide mapping in the future with the development of models and the extension of geological data related to submarine landslides.

Keywords: submarine landslide; machine learning; hazard susceptibility; spatial distribution

Citation: Du, X.; Sun, Y.; Song, Y.; Xiu, Z.; Su, Z. Submarine Landslide Susceptibility and Spatial Distribution Using Different Unsupervised Machine Learning Models. *Appl. Sci.* **2022**, *12*, 10544. https://doi.org/10.3390/app122010544

Academic Editors: Yuzhu Wang, Jinrong Jiang and Yangang Wang

Received: 1 September 2022
Accepted: 14 October 2022
Published: 19 October 2022

1. Introduction

A submarine landslide is a destructive phenomenon of marine geological disasters. Large-scale submarine landslides can even cause long-distance migration of thousands of cubic kilometers of sediment [1], damage various engineering facilities such as offshore oil production platforms [2] and submarine optical cables [3], and even cause tsunamis [4]. This can result in a number of incidents, such as communication failure and platform collapse, which pose a significant risk to human life and property. Therefore, there is a pressing need for submarine landslide stability before engaging in offshore engineering activities.

At present, the main research directions of submarine landslides include: using high-precision geophysical detection to identify and classify landslide morphology [5,6], carrying out stability calculations of submarine landslides by the numerical analysis method [7–9], and simulating the landslide process through physical model tests such as a conventional water tank or centrifuge [10,11]. Despite attempts through the above traditional studies, the research on risk assessment and categorization is still insufficient due to the complicated control circumstances of submarine landslides, the large number of trigger factors, and the difficulty of monitoring.

Machine learning and deep learning techniques have been proven to be powerful and promising tools in many geotechnical applications [12–16]. Chen et al. [12] designed landslide spatial models using maximum entropy, support vector machine, and artificial neural network methods. Tse et al. [15] performed an unsupervised learning approach to study the synchroneity of past events in the South China Sea. Qi and Tang [16] used integrated metaheuristic and machine learning approaches for slope stability prediction. Deep learning convolutional neural networks [17] and support vector machines [18] are also used in landslide detection. Even though the mentioned methods performed well for landslide modeling in a given area, there is no conclusive information about which model is the best for other regions. In addition, the application of the recently developed techniques and methods for a more accurate evaluation of the predictive capability of landslide susceptibility models should be evaluated further.. At present, the main research objects of landslides using machine learning are landslides on land, but few types of research are on submarine landslides. The problem of zoning submarine landslide hazards is still a difficult area for landslide research. On the other hand, machine learning excels at resolving nonlinear problems without the need of explicit mathematical relationships. Therefore, it is essential to investigate if machine learning algorithms can be utilized for zoning undersea landslide hazards and to research how well various machine learning methods perform.

The primary purpose of this study is to offer an integrated strategy for assessing submarine landslide susceptibility that uses unsupervised machine learning models to evaluate landslide risk and partition the affected region. Submarine landslides in the Yellow River Estuary are selected for study and validation of the suggested method. Three machine learning models based on k-means, spectral clustering, and hierarchical clustering are developed and compared for performance evaluation.

2. Study Area Description

The Yellow River Estuary is located in the north of Shandong Peninsula, China (118.73° E–119.65° E, 38.1° N–38.3° N) (Figure 1). In this sea region, the water depth ranges from 0 to 18 m. When the water depth is less than 15 m, the predominant sediment type is silt combined with a minor quantity of silty sand; when the water depth is greater than 15 m, the predominant sediment type is silty clay. The wave height increases gradually with the increase in water depth and reaches the maximum value near the water depth of 10 m. As a consequence, wave forces on the seabed increase first, and then decrease with the increase in water depth. The seabed gradient ranges from 1/2000 to 1/500. Wave-induced liquefaction, which can cause seabed sediment to lose stability and slide, is the main geological disaster in the sea area.

There are two main factors for the selection of the Yellow River Estuary as the study area. First, the detailed geological data of the study area, such as water depth, sediment types, waves, currents, and so on, have been collected by the First Institute of Oceanography, MNR, and can be used in this study. It is of vital importance to obtain detailed geological environment data, as the acquisition of data about submarine landslides is very difficult and incomplete. Second, there are a great many submarine landslides and human activities located in this sea area, which means this study has research significance and practical engineering safety-guidance significance. There are a lot of micro-submarine landslides that have been discovered since the 1990s in the study area. The results of the field geophysical investigation [19] show that the form factor of the submerged delta landslide zone in the Yellow River Estuary is mainly hydrodynamically triggered by submarine liquefaction. Furthermore, the SINOPEC Shengli Oilfield is in this sea area, and as a consequence, hundreds of submarine oil platforms and submarine cables are located in this area.

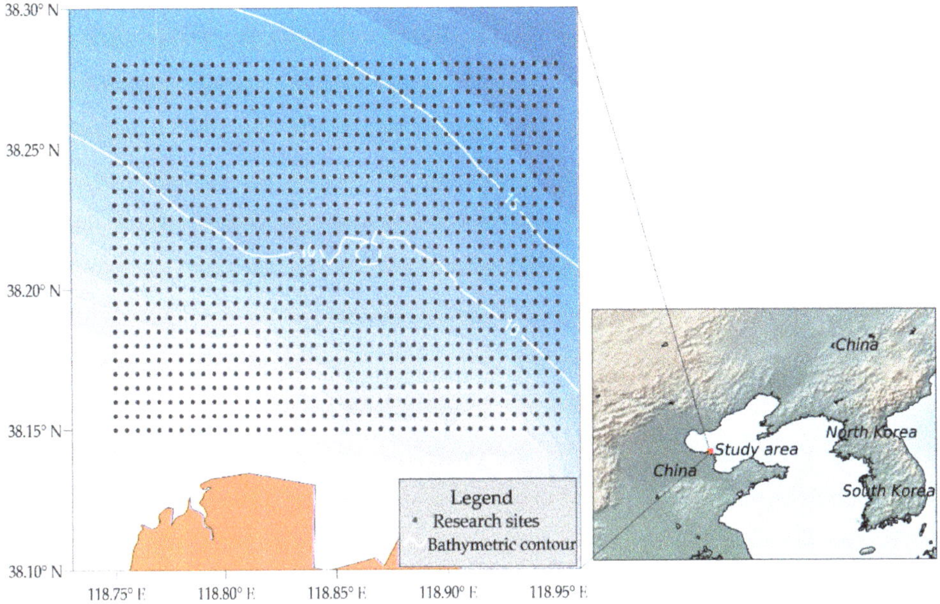

Figure 1. Location map of the study area.

3. Data and Methods

3.1. Overall Workflow

The main objective of this paper is to use the data of various types of submarine-landslide-hazard impact factors as the basis for regional classification of submarine landslide hazards after data preprocessing and machine learning modeling. The study in this paper can provide an exploration of the hazard classification of global submarine landslides. The workflow of this study is shown in Figure 2, and includes the following specific steps:

- Data collection.

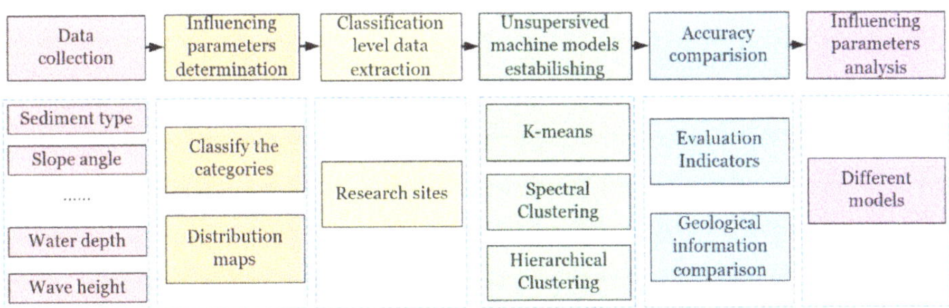

Figure 2. The general concept of the research method.

The data of this study are obtained from long-term geophysical surveys and in situ monitoring in the study area by the First Institution of Oceanography, MNR [20]. After data collection, we should determine suitable geological factors as submarine-landslide-influencing parameters. Suitable impact parameters should be able to cover both the geological, hydrological, and human influential factors of landslides and be more readily available as common parameters.

- Influencing parameter determination.

After determining the factors influencing the submarine landslides, each data point was interpolated separately so that the values of all influencing factors were included at each study site in the study area. Subsequently, the data were classified into several categories according to the characteristics of different influencing factors, and the category distribution maps were drawn.

- Classification-level data extraction.

The coordinate values of the study-point locations are defined and the categories of each factor are extracted from the impact-factor classification map obtained in the previous step. The coordinate values of the study-point locations are defined and the categories of each factor are extracted from the impact-factor classification map obtained in the previous step. Thus, each study point corresponds to multiple classes of impact-factor parameters that can be used in the next step of unsupervised machine learning model training.

- Establishing unsupervised machine learning models.

Establish 3 unsupervised machine learning models using k-means, spectral clustering, and hierarchical clustering. The different parameters in the model are first modeled several times, and subsequently, the model parameter with the best prediction is selected to build the final established model.

- Accuracy comparison.

Compare the accuracy and rationality of different predicting results and choose the best one. Both the mathematical test metrics and the measured geological conditions should be used to test the models' accuracy. The Calinski–Harabasz index, silhouette index, and Davies–Bouldin index are used to calculate the mathematical accuracies. The liquefication zonation is used to calculate the geological rationality.

- Influencing parameter analysis.

Study the importance of all the landslide-influencing parameters by excluding them individually using the best model and test the accuracies with evaluating indicators.

3.2. Landslide-Influencing Parameters

It is very important to choose the suitable influencing factors for submarine landslide assessment. There is no absolute standard parameter when classifying the hazards of submarine landslides. This issue remains one of the difficult problems in the field of research on submarine landslides. The reason is that there are too many influencing factors and it is difficult to obtain the corresponding parameters. From the point of view of geological analysis, geological factors, hydrodynamic factors, topographic factors, and human activities should be taken into account. As much information as possible should be collected to satisfy these four requirements.

In this study, we have selected carefully out of all the various choices available based on the nature of submarine landslide occurrences concerning the characteristics of geology, hydrology, geomorphology, and the impact of human engineering activities. Therefore, 9 factors were selected, namely, sediment type, slope, soil strength, water depth, wave height, maximum current velocity of the bottom, liquefaction, erosion, and human engineering activities (Figure 3). The research data of the 9 factors were obtained by the First Institute of Oceanography, MNR, China through geophysical sounding, drilling, and monitoring surveys in the Chengdao sea area of the Yellow River Estuary [18], and contain detailed information on various geological features of the study area. Each factor was divided into 3 or 4 classes based on the range of data, geological background, and experts' experience in this study area. At last, 1107 points, of which longitudes vary from 118.75° N to 118.95° N, latitudes change from 38.15° N to 38.28° N, and 0.05 degrees is the interval, were selected as the research sites (Figure 1). All the data used in this study were collected from projects of the First Institute of Oceanography, MNR.

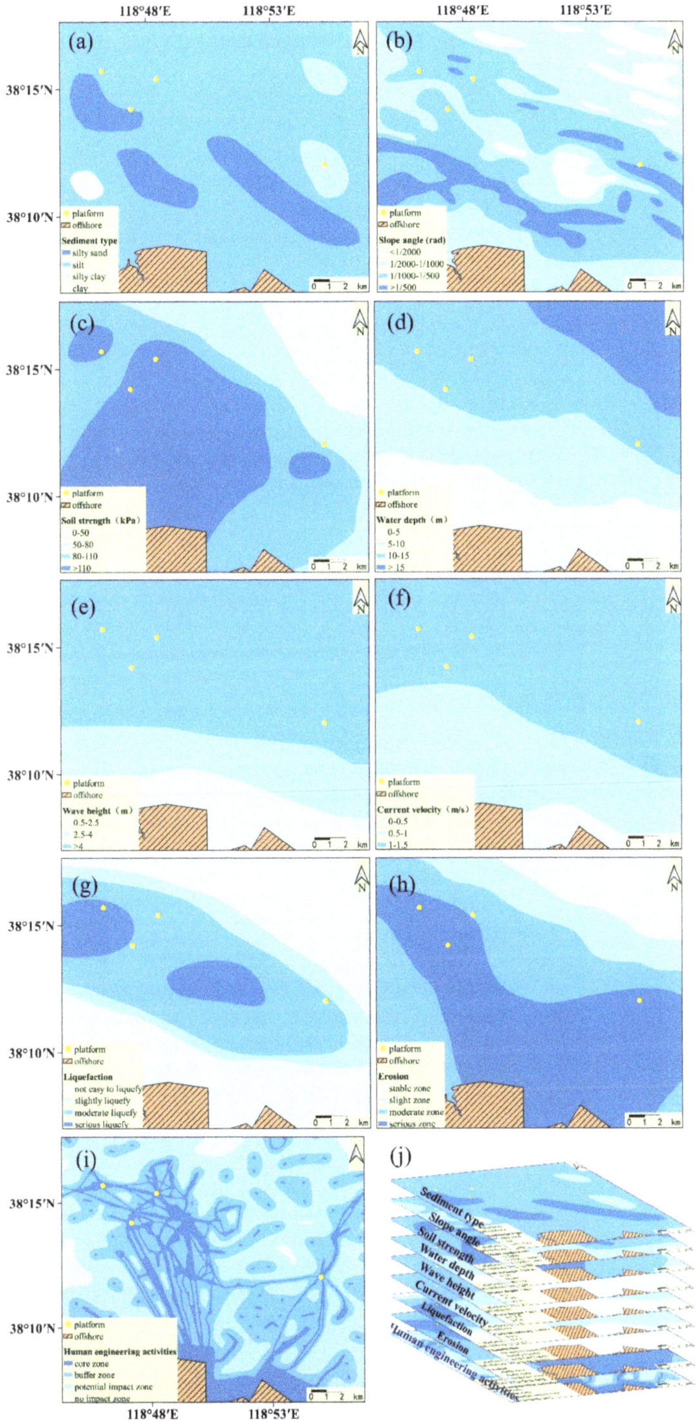

Figure 3. Location map of the study area. (**a**) Sediment type; (**b**) slope angle; (**c**) soil strength; (**d**) water depth; (**e**) wave height; (**f**) current velocity; (**g**) liquefaction; (**h**) erosion; (**i**) human engineering activities; (**j**) overlay map.

Sediment type plays an important role in the study of submarine landslide susceptibility, as different types of sediments have different physical and mechanical properties, which can affect the difficulty of geological disasters. Studies have shown that sediment type has a large influence on landslide stability [21]. In this study area, sediments are divided into 4 classes: silty sand, silt, silty clay, and clay.

Slope angle is a significant factor in the development of submarine landslide susceptibility and the angles were calculated by the change in water depth. Slope angles are subdivided into 4 categories: <1/2000 radian, 1/2000–1/1000 radian, 1/1000–1/500 radian, and >1/500 radian. The places with large sea-bottom slopes are mainly located at 6 m and 10 m water-depth contours.

Soil strength affects the stability and sliding difficulty of the submarine landslide. The greater the soil strength, the harder the slide occurs. Soil strength is divided into 0–50 kPa, 50–80 kPa, 80–110 kPa, and >110 kPa. The classification of soil strength is mainly based on data from boreholes in this study area.

Water depth, which was obtained by single-beam and multibeam bathymetric instruments, can influence the strength of waves acting on the seabed. It can be classified into 4 classes: 0–5 m, 5–10 m, 10–15 m, and >15 m.

Wave height is a very important factor because it represents the energy that a wave contains. It was collected by pressure wave and tide gauges. The wave height increases first with the depth of water, but there is no obvious increase after reaching a 9 m depth. The study area can be divided into 3 classes, which are 0.5–2.5 m, 2.5–4 m, and >4 m.

The maximum current velocity of the bottom determines the shear stress of the current on the seabed, which may cause erosion. Current velocity increases as the water depth increases and can be classified into 3 classes: 0–0.5 m/s, 0.5–1 m/s, and 1–1.5 m/s.

Liquefaction is the most serious geological hazard in the Yellow River Estuary. There are hundreds of liquefaction zones in the study area, which were discovered by geophysical explorations. Liquefaction zones are mainly distributed between a 6 m to 12 m water depth, where the strength of hydrodynamic action is the strongest. Liquefaction is divided into 4 classes: not easy to liquefy (liquefaction depth < 0.5 m), slightly liquefy (0.5 m < liquefaction depth < 2 m), moderate liquefy (2 m < liquefaction depth < 4 m), and serious liquefy (liquefaction depth > 4 m). Seabed sediments are easy to slide after liquefaction as their bearing capacity reduces greatly.

Erosion is divided into 4 classes, which are the stable zone (<0.02 m/s), slight zone (0.02–0.05 m/s), moderate zone (0.05–0.1 m/s), and serious zone (0.1 m/s). The serious zone and moderate zone are mainly distributed in a water depth of less than 12 m.

Human engineering activities are mainly offshore production platforms, submarine pipelines, cables, and so on in this study area. They can be divided into 4 categories: core zone, buffer zone, potential-impact zone, and no-impact zone. The actual scope of various engineering structures is named the core zone. The buffer zone is the core area extending 500 m outwards. The potential-impact zone is where the buffer zone extends another kilometer outward, and other areas are the no-impact zones.

3.3. *Unsupervised Machine Learning Models*

Unsupervised machine learning is a major part of machine learning. It can study the intrinsic relationship of datasets without data labels when dealing with practical problems. The main applications of unsupervised learning are: segmenting a dataset by some shared attributes; detecting exceptions that are not suitable for any group; and simplifying datasets by aggregating variables with similar properties. Among these, the objective of this study is to explore the susceptibility of submarine landslides. As a consequence, an important unsupervised machine learning class named clustering, which contains k-means, spectral, and hierarchical clustering, was selected as the study method. Each clustering method was built after parameter selection with internal validation measures: the Calinski–Harabasz index [22], silhouette index [23], and Davies–Bouldin index [24]. The more precise the clustering result, the higher the Calinski–Harabasz score and silhouette index. The lower the Davies–Bouldin index, the more accurate the clustering result. The clustering results' performances were validated with external validation measures; for instance, liquefaction distribution, hydrodynamic action, slope angle, etc.

3.3.1. k-Means

The k-means clustering algorithm is a typical unsupervised machine learning model that is widely used for the clustering analysis [25] of non-labeled data. The advantage of k-means is that it is easy to implement and visualize the result. The number of clusters is the only parameter that needs to be specified beforehand. To build a k-means model:

(a) Predetermine the clustering number k.
(b) The k-clustering prime points are randomly selected as $\mu_1, \mu_2, \ldots, \mu_k$.
(c) All the points are assigned to the nearest centroid and clusters are formed. Calculate the distances between every point to the centroid in each cluster.

$$dist(x, \mu) = \sqrt{\sum_{i=1}^{n} (x_i - \mu_i)^2}$$

where x is the sample point; μ is the center of mass of the cluster; n is the number of features in each sample point; and i is each feature of the constituent point x.
(d) Summarize the total distances of all clusters:

$$Cluster\ Sum\ of\ Square(CSS) = \sum_{j=0}^{m} \sum_{i=0}^{n} (x_i - \mu_i)^2$$

$$Total\ Cluster\ Sum\ of\ Square = \sum_{l=1}^{k} CSS_l$$

where m is the number of samples in a cluster and j is the number of each sample.
(e) Calculate the minimum quadratic error from the data point to the center of each cluster, and move the center to the point.
(f) Repeat the calculation from step (c) until the total cluster sum of squares does not change or reach the maximum iteration times.

3.3.2. Spectral Clustering

Spectral clustering is another unsupervised machine learning model, which clusters through the characteristic vector of the Laplacian matrix of sample data. Spectral clustering maps data from a high-dimensional space to low-dimensional space, and then, uses other clustering algorithms to cluster in a low-dimensional space. Compared with k-means, spectral clustering uses a dimension reduction algorithm, which is more suitable for high-dimensional data processing and more effective for sparse data processing. Spectral clustering outputs clusters $A_1, A_2, \ldots, A_n$ by inputting n sample points $X = \{x_1, x_2, \ldots, x_n\}$ and the number of clusters k. In this model, the kernel function parameter and cluster number are the influential parameters. The specific steps are below:

(a) Calculate the similarity matrix W of $n * n$, which includes the minimum proximity method, k-proximity method, and full-connection method. The full-connection method used in this study is as described:

$$s_{ij} = s(x_i, x_j) = \sum_{i=1, j=1}^{n} \exp \frac{-\|x_i - x_j\|^2}{2\sigma}$$

where s_{ij} = the similarity matrix and σ = kernel function parameter, which controls the neighborhood width of the sample point.
(b) Calculation matrix D:

$$d_i = \sum_{j=1}^{n} \omega_{ij}$$

where D is the $n * n$ diagonal matrix formed with d_i.

(c) Calculate the Laplacian matrix $L = D - W$.

(d) Calculate the characteristic value of D and sort it from small to large, then take the first k characteristic values and calculate their feature vector $u_1, u_2, \ldots, u_n$.

(e) Form the matrix $U = \{u_1, u_2, \ldots, u_n\}, U \in R^{n+k}$.

(f) Let $y_i \in R_k$ be the vector of the line i of $U, i = 1, 2, \ldots, n$.

(g) Cluster the datasets $Y = \{y_1, y_2, \ldots, y_n\}$ into clusters $C_1, C_2, \ldots, C_k$.

(h) Output clusters $A_1, A_2, \ldots, A_k$, among which $A_i = \{j|y_i \in C_i\}$.

3.3.3. Hierarchical Clustering

Hierarchical clustering is another unsupervised algorithm that is based on hierarchical methods. When using hierarchical clustering, each object is regarded as a cluster, and then, the clusters are merged step by step according to some rules so that the number of cluster classes is reached. The advantages are: the similarity of distance and rule is easy to define and is limited, and the hierarchical relationship of classes can be found and can be clustered into other shapes. Meanwhile, the disadvantages are: the computational complexity is too high, a singular value can also have a great influence, and the algorithm is likely to cluster into chains. To build a hierarchical clustering model:

(a) Each object is regarded as a class, and the minimum distance between two objects is calculated;

(b) The two classes with the smallest distance are combined into a new class;

(c) Recalculate the distance between the new class and all classes;

(d) Repeat (a) and (b) until all classes are finally merged into one class.

The effective parameters are cluster number, linkage, and affinity, of which linkage contains Ward, average, and complete; and affinity contains Euclidean, Manhattan, and cosine. Ward can only be combined with Euclidean when averaged, which provides the best performance but a large computation, and can be combined with Manhattan and cosine. We should study the parameters and validate the result before building the final hierarchical clustering model.

3.4. *Validation*

Cluster results of k-means, spectral clustering, and hierarchical clustering validated with different measures are shown in Figure 4. It can be seen from (1)~(6) and (10)~(12) that all the three cluster methods performed best when the cluster number is 2, and decreased when the cluster number increased. In the mathematics view, the data should be divided into two classes as they performed best. However, we should consider our geological needs as an important classification factor as well. This sea area can be divided into two classes, with the seabed sediments mutating at about a 14 m water depth.

Nevertheless, we need a more accurate range of submarine landslide susceptibility with acceptable mathematical accuracy. The cluster number of 3 is still insufficient to classify the study region because the classification result may be easily inferred from the hydrodynamic characteristics. As a result, the final cluster number is 4, which corresponds to the classifications very high, high, low, and very low.

As seen from (7) to (9), the clustering result performed best when gamma was equal to 0.01 with the Calinski–Harabasz index and gamma equal to about 1 using the silhouette index and Davies–Bouldin index. Therefore, the kernel function parameter gamma is 1. It is shown in (13)~(15) that the result performed best when the affinity method is Manhattan. The specific parameters used in the three cluster models can be seen in Table 1.

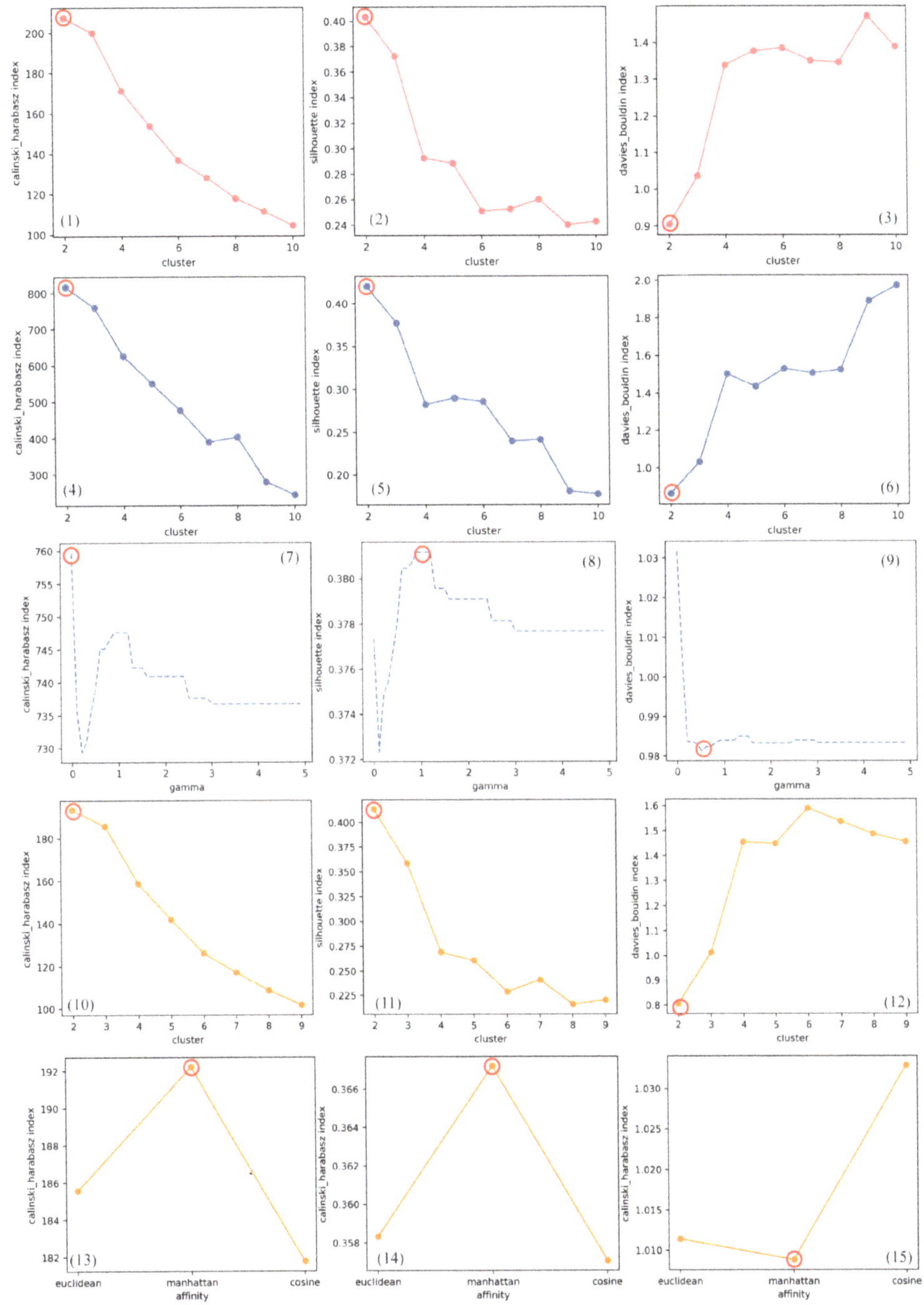

Figure 4. Cluster results of k-means, spectral clustering, and hierarchical clustering were validated with the Calinski–Harabasz index, silhouette index, and Davies–Bouldin index. (1)~(3) Validate cluster numbers of k-means. (4)~(6) Validate cluster numbers of spectral clustering. (7)~(9) Validate kernel function parameter of spectral clustering. (10)~(12) Validate cluster numbers of hierarchical clustering. (13)~(15) Validate affinity of hierarchical clustering. The red circle indicates the location where the model achieved the best prediction.

Table 1. Parameters used in different models.

Model	Cluster Number	Gamma	Affinity	Linkage
k-means	4	None	None	None
Spectral	4	1	None	None
Hierarchical	4	None	Manhattan	Average

The parameters are not necessary for the model when it shows "None".

All machine learning calculations in the article were carried out using the scikit-learn machine learning package [26], an open-source python library, via a laptop with windows 10, 16G RAM, CPU R 5800H, and GPU 3060.

4. Results

This section outlines the clustering results of the submarine landslide susceptibility using k-means, spectral, and hierarchical models after parameter validation and selection. The study area was divided into four areas of different submarine landslide susceptibilities without real geological labels. To define the final labels, all the geological parameters should be considered. As mentioned in Section 3.2, the hydrodynamic force to the seabed increases first with the deepening of water depth and then decreases. Moreover, the sediment type changes from silt to clay, which is difficult to be influenced, when the water depth is deeper than 15 m. Therefore, the submarine landslide susceptibility labeling principle of this area is that: the least serious part is the clay region; the second least serious part is the shallow water region; the most serious part is around the 10 m bathymetric contour; and the second most serious part is beside or around it.

As shown in Figure 5, the study area was divided into four parts of submarine landslide susceptibility using the k-means model. The very high-susceptibility part is located at a water depth of 5–11 m, and the edge is a discrete distribution. The high-susceptibility part is distributed at a water depth of 11–13 m. The low-susceptibility part is situated at a water depth of less than 5 m. The very low-susceptibility part is located at a water depth deeper than 13 m.

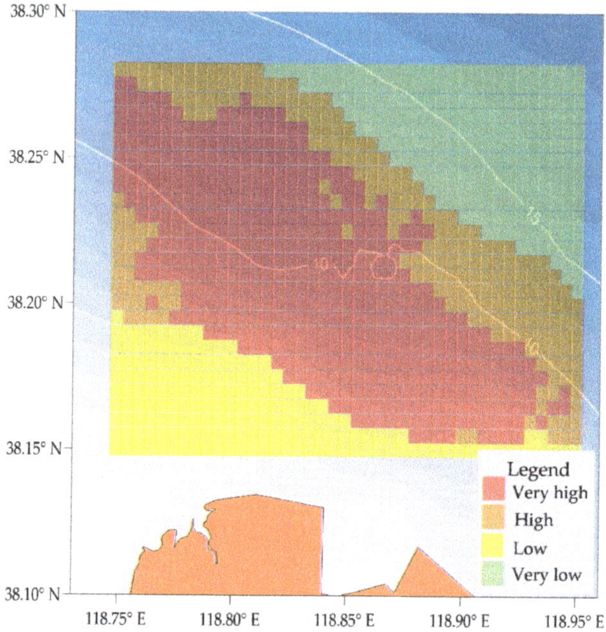

Figure 5. Distribution of submarine landslide susceptibility using k-means. The different colors represent different submarine landslide susceptibilities.

It can be seen from Figure 5 that the study area was divided into four parts of submarine landslide susceptibility using the spectral clustering model. The very high-susceptibility part is located at a water depth of 5–12 m. The high-susceptibility part is distributed at a water depth of 12–15 m. The low-susceptibility part is situated at a water depth of less than 5 m. The very low-susceptibility part is located at a water depth deeper than 15 m. Compared with the results obtained by k-means, the distribution of submarine landslide susceptibility using spectral clustering is more continuous than that using the k-means algorithm. Furthermore, the very high-susceptibility part is wider and the very low part is narrower in Figure 6 than in Figure 5.

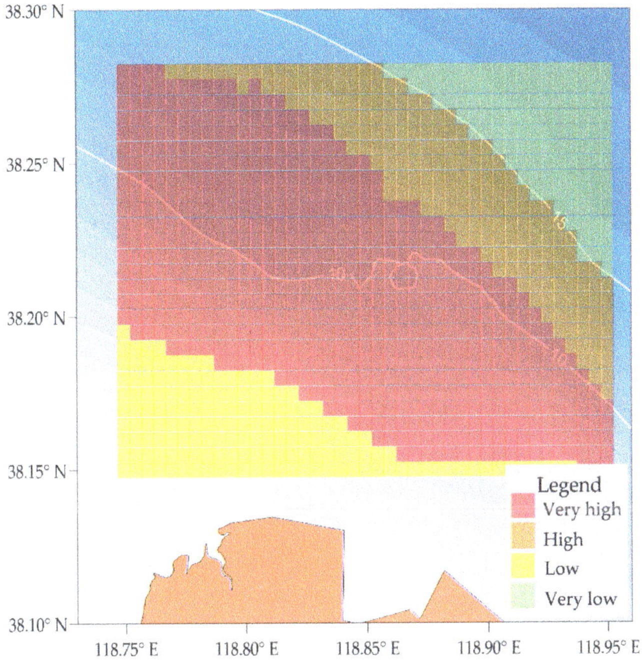

Figure 6. Distribution of submarine landslide susceptibility using spectral clustering.

As shown in Figure 7, the study area was divided into four parts of submarine landslide susceptibility using the hierarchical clustering model. The very high-susceptibility part is located at a water depth of 5–8 m. The high-susceptibility part is distributed at a water depth of 8–13 m. The low-susceptibility part is situated at a water depth of less than 5 m, and the very low-susceptibility part is located at a water depth deeper than 13 m. By comparing the results of the three algorithms, we can find that the very high-susceptibility part obtained by using the hierarchical clustering model is much less than other two of the other methods.

Generally, the low- and very low-susceptibility parts of these three methods are very close. The main differences are in the size and distribution of very high and high areas. The three unsupervised machine learning methods obtained two main common parts: one is a low-hazard region at shallow depths of 5 m; the other is a very low-hazard region at depths of 13 m. The reason for this phenomenon is that the hydrodynamic conditions are relatively weak in these two parts of the area, and the influence of various geological-impact parameters on submarine landslides is also small; thus, the results obtained by different algorithms are more consistent.

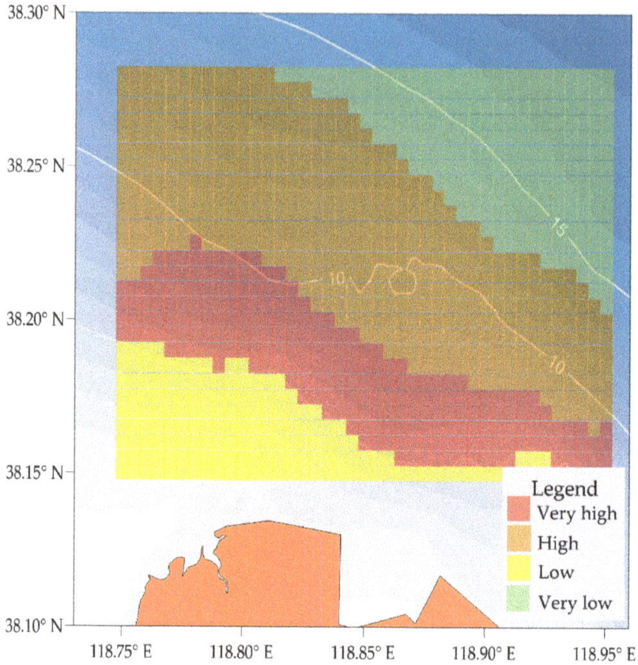

Figure 7. Distribution of submarine landslide susceptibility using hierarchical clustering.

5. Discussion

5.1. Model Performance Comparison

To test the performance of cluster results, both internal validation measures and external validation measures were used. For internal validation measures, we validated the accuracy of three different submarine landslide susceptibility models by using the Calinski–Harabasz index, silhouette index, and Davies–Bouldin index. The three indexes use different algorithms to examine the merits of the classification results from a mathematical perspective. As shown in Figure 8, the k-means model performed best under the evaluation of the Calinski–Harabasz index, whereas the spectral clustering model performed best when evaluating the silhouette index and Davies–Bouldin index. Therefore, spectral clustering has the best performance compared to k-means and hierarchical clustering in the internal validation measures.

In general, external causes of submarine landslides include earthquakes, gas hydration, wave action, volcanic activity, tsunamis, etc. There are no severe geological phenomena such as earthquakes, tsunamis, and volcanoes in the study area. The main external influence factor is the liquefaction of seabed soil caused by waves. Therefore, for the external validation measure, the three distributions of submarine landslide susceptibility were compared with the distribution of liquefaction depth (Figure 9). As shown in Figure 9, the area with the deepest liquefaction depth (ellipse A) is distributed at a water depth of 6–12 m. The area of ellipse A agrees well with the submarine landslide susceptibility results using k-means and spectral clustering, whereas the result obtained using hierarchical clustering is quite different from ellipse A. The area with a deep liquefaction depth (ellipse B) is distributed at a water depth of 12–15 m. It is in good agreement with the result obtained using hierarchical clustering, and the areas obtained by k-means and hierarchical clustering are different from the area of ellipse B. The area with a small, shallow liquefaction depth (ellipse C) is distributed at a water depth of less than 5 m, which agrees well with all the clustering results. The area with a very shallow liquefaction depth (ellipse D) is distributed at a water depth deeper than 15 m. It is in good agreement with the result obtained using hierarchical clustering, and the areas obtained by k-means and hierarchical clustering are larger than the area of ellipse D. As a result, the submarine landslide susceptibility results using spectral clustering performed better than those obtained using k-means and hierarchical clustering in the external validation.

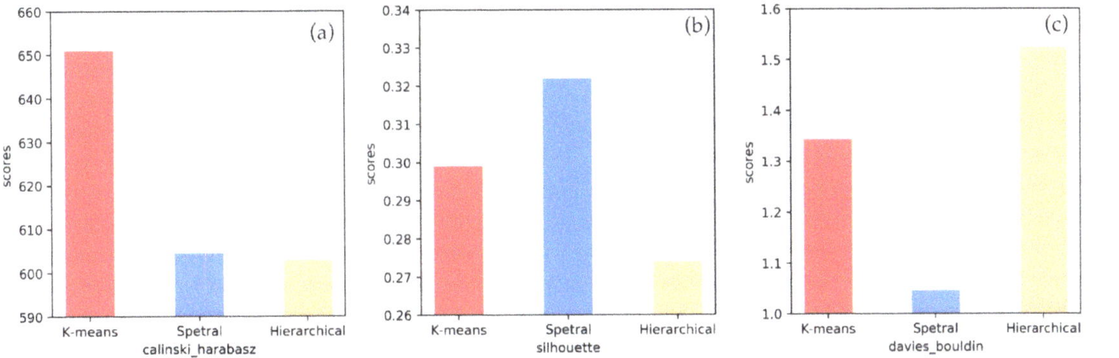

Figure 8. Comparison of clustering results validated with different internal validation measures. (**a**) Performance of different models under Calinski–Harabasz validation methods. (**b**) Performance of different models under silhouette validation methods. (**c**) Performance of different models under Davies–Bouldin validation methods. The higher the Calinski–Harabasz index and silhouette index are, the more accurate the clustering result will be. The lower the Davies–Bouldin index is, the more accurate the clustering result will be.

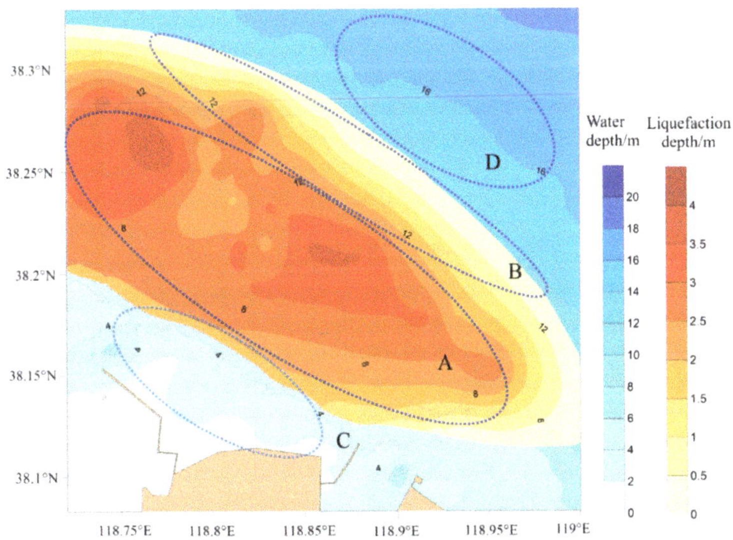

Figure 9. Distribution of liquefaction depth in the Yellow River Estuary. The area of ellipses A, B, C, and D represents the very deep, deep, shallow, and very shallow parts of liquefaction depth, respectively.

In conclusion, all three models used in this study are capable of producing correct clustering results, with the spectral clustering model being the most precise when grouping the undersea landslide susceptibility. Although the best algorithm can be obtained by internal verification methods, the scores of different algorithms are not different enough to represent the difference. To obtain the best results, it is still necessary to combine external verification methods at the same time.

5.2. Comparison of Model Results with Other Studies

In order to verify the accuracy of the model calculations in this paper, comparisons with the results of other studies are still needed. Therefore, we selected the results of a geophysical survey and the results of traditional GIS-based landslide analysis methods to compare and analyze with the conclusions of this paper.

In the 1980s, a comprehensive, integrated geophysical survey was conducted in the Yellow River Estuary waters [19]. The results of the survey showed that a large number of microslides on the leading edge of the delta existed at a water depth between 4 and 12 m in the study area. It can be seen from Figure 5 that the classification of landslides as hazard results using the spectral clustering algorithm shows that the areas with very high hazards are concentrated at a water depth range of 5–12 m. The simulation results are highly consistent with the actual survey results, indicating the accuracy and reliability of the algorithm in this paper.

As for the conventional approach used to analyze submarine landslide hazards, the GIS-based analyzing method was also used to study the stability of submarine landslides in the Yellow River Estuary [27]. Results show that the most prone landslide areas in the study area are located at a water depth between 8 and 13 m, and the more prone landslide areas are located at a water depth between 5 and 15 m, with an average water depth of 10 m. The range of landslide-hazard areas derived from GIS results is consistent with the overall distribution range and trend compared with those derived in this paper, and the range of landslide hazards is slightly smaller. The model calculation results in this paper are closer to the results of the actual geophysical survey compared with the GIS results.

Therefore, it is clear that the unsupervised machine learning method used in this paper has high reliability and stability after being compared with the traditional methods of submarine-landslide-hazard analysis and the actual geophysical survey results in the field.

Although just one research area was used for the study of submarine landslide hazards, the general geological formation conditions and triggering mechanisms of submarine landslides in all study regions are comparable, despite the differences in causes and triggering variables. As a result, this paper's research technique and research hypothesis can serve as a guide for the global study of submarine landslide hazards.

5.3. Importance of Landslide-Influencing Factors

To figure out the significance of influencing elements, each factor was eliminated and the results were recalculated. New cluster results with different factors were compared with the original clustering results using the Calinski–Harabasz index, silhouette index, and Davies–Bouldin index. Nine clustering results correspond to the missing influence factors and the evaluation scores were normalized. The higher the normalized ratio is, the less important this factor is. The order of importance in this study does not represent the absolute order of the corresponding impact factors, but only represents the relative order in the study area.

Test results can be seen in Figure 10, where the CH represents the normalized ratio of the Calinski–Harabasz index and the SI means the normalized ratio of the silhouette index. The DAV is obtained by 2 minus the normalized ratio of the Davies–Bouldin index so that the trend is the same as the other two indexes. As seen in Figure 10, the model without liquefaction obtained the lowest score, which means liquefaction is the most important factor that influences landslides in the study area. Models that exclude water depth, wave height, and soil strength obtained higher scores than those that exclude liquefaction, which means they are the second most important factors. These three elements may affect whether liquefaction happens, but none of them can predict it on their own. Consequently, these components are less significant than liquefaction, but more significant than other factors. As for sediment type, erosion, and maximum current velocity of the bottom, they are less important than the factors mentioned before. There is a great correlation between sediment types and soil strength. However, the physical and mechanical properties of the same sediment may be different because of the different depositional states and times.

Therefore, soil strength has a greater impact on landslides. Erosion and maximum current velocity of the bottom have certain influences on the stability and strength of sediments, but are not decisive; therefore, the importance degree is relatively low. The least important factors are slope angle and human engineering activities in this study area. Slope angle has an important influence on the landslide according to previous studies [28,29], but the difference in slope angle in this study area is not large enough, and thus, the influence degree is low. As described in Section 3.2, the biggest slope angles are >1/500 radian, which is about equal to 0.11°. However, Hance [30] counted 399 cases of submarine landslide slope angles and found that the most frequent value was 3~4° and the average value was 5.8°. These data are far larger than the slope value in our study area, and the average value is 50 times the maximum value in the region; thus, the slope angle is negligible and is very unimportant in the landslide evaluation of the study area. As for human engineering activities, some engineering activities, such as hydrate exploitation [31,32], can play an important role in the formation of submarine landslides, but the engineering activities in the study area are mainly offshore platforms and submarine pipelines, which are widely distributed, and therefore, their importance is very low and can be ignored.

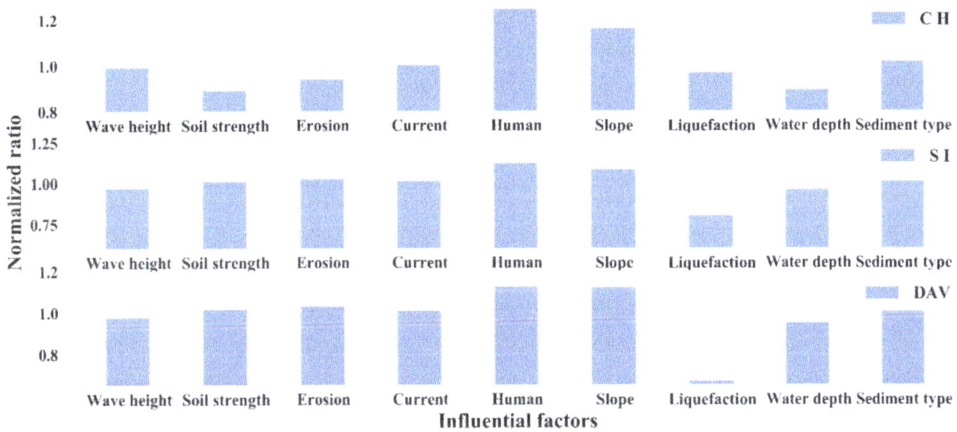

Figure 10. The normalized score of clustering results changed with different influence parameters. The normalized results were obtained by dividing the new scores of eight parameters by the scores of nine parameters.

In conclusion, various influencing variables have varying degrees of impact influence on submarine landslide risk assessment, and the significance relies on the degree of correlation between the landslide and its distribution in the study area. The order of importance and degree of effect of variables acquired in this research only reflect one study area; the techniques described in Section 5.2 must be employed to conduct particular analyses in other study areas.

6. Conclusions

In this paper, a submarine landslide susceptibility assessment was carried out using unsupervised machine learning models in the Yellow River Estuary, China. Nine influential factors were selected to analyze the susceptibility of submarine landslides based on terrain data and remote sensing images. We used different unsupervised machine learning models to classify landslide risk and discussed the accuracy of the model and the importance of a single factor. The main conclusions are as follows:

(1) Unsupervised machine learning models can be used to study and assess submarine landslide susceptibility and provide high accuracy.

(2) Results using the spectral clustering method have the highest accuracy among k-means, spectral clustering, and hierarchical clustering after testing with both internal validation measures and external validation measures.
(3) In this study area, the order of importance of submarine-landslide-influencing factors is as follows: liquefaction, water depth, wave height, soil strength, sediment type, erosion, maximum current velocity of the bottom, slope angle, and human engineering activities. In different research areas, the importance of each impact factor is different, which needs specific analysis.

Due to the complexity of the elements impacting the submarine-landslide-hazard triggers and the difficulties of monitoring submarine landslide sites, it is challenging to determine the precise location and hazard information of each landslide in the research region. Currently, unsupervised machine learning can only be conducted using limited data for semiquantitative description. For a more in-depth examination of submarine landslide hazards, accurate training data is necessary. In the future, we must, therefore, place a greater emphasis on the collection of field data for undersea landslide identification and monitoring.

Author Contributions: Conceptualization, Y.S. (Yongfu Sun); methodology, X.D.; program, X.D.; validation, Y.S. (Yupeng Song) and Z.X.; resources, Z.S.; data curation, Z.S.; writing—original draft preparation, X.D.; writing—review and editing, X.D.; visualization, X.D.; supervision, Y.S. (Yupeng Song); project administration, X.D. and Y.S. (Yongfu Sun); funding acquisition, X.D. All authors have read and agreed to the published version of the manuscript.

Funding: This research was funded by the Foundation item: The National Natural Science Foundation of China under contract NO. 42102326; the Basic Scientific Fund for National Public Research Institutes of China under contract NO. GY0222Q05; and The Shandong Provincial Natural Science Foundation, China under contract NO. ZR2020QD073.

Institutional Review Board Statement: Not applicable.

Informed Consent Statement: Not applicable.

Data Availability Statement: Data that support the findings of this study are available from the corresponding author upon reasonable request.

Conflicts of Interest: The authors declare no conflict of interest.

References

1. Micallef, A.; Masson, D.G.; Berndt, C.; Staw, D. Morphology and mechanics of submarine spreading: A case study from the Storegga Slide. *J. Geophys. Res.* **2007**, *112*, 739. [CrossRef]
2. Liu, J.; Tian, J.; Ping, Y. Impact forces of submarine landslides on offshore pipelines. *Ocean Eng.* **2015**, *95*, 116–127. [CrossRef]
3. Mosher, D.C.; Monahan, P.A.; Barrie, J.V.; Courtney, R.C. Coastal Submarine Failures in the Strait of Georgia, British Columbia: Landslides of the 1946 Vancouver Island Earthquake. *J. Coast. Res.* **2004**, *20*, 277–291. [CrossRef]
4. Michael, A.F.; William, R.N.; Greene, H.G.; Homa, J.L.; Ray, W.S. Geology and tsunamigenic potential of submarine landslides in Santva Barbara Channel, Southern California. *Mar. Geol.* **2005**, *224*, 1–22. [CrossRef]
5. Wang, W.; Wang, D.; Wu, S.; Völker, D.; Zeng, H.; Cai, G.; Li, Q. Submarine landslides on the north continental slope of the South China Sea. *J. Ocean Univ. China* **2018**, *17*, 83–100. [CrossRef]
6. Ilstad, T.; De Blasio, F.V.; Elverhøi, A.; Harbitz, C.B.; Engvik, L.; Longva, O.; Marr, J.G. On the frontal dynamics and morphology of submarine debris flows. *Mar. Geol.* **2004**, *213*, 481–497. [CrossRef]
7. El-Ramly, H.; Morgenstern, N.R.; Cruden, D.M. Probabilistic slope stability analysis for practice. *Can. Geotech. J.* **2002**, *39*, 665–683. [CrossRef]
8. Griffiths, D.V.; Lane, P.A. Slope stability analysis by finite elements. *Géotechnique* **1999**, *49*, 387–403. [CrossRef]
9. Ijaz, N.; Ye, W.; Rehman, Z.; Dai, F.; Ijaz, Z. Numerical Study on Stability of Lignosulphonate-Based Stabilized Surficial Layer of Unsaturated Expansive Soil Slope Considering Hydro-Mechanical Effect. *Transp. Geotech.* **2022**, *32*, 100697. [CrossRef]
10. Bradshaw, A.S.; Tappin, D.R.; Rugg, D.A. The Kinematics of a Debris Avalanche on the Sumatra Margin. *Int. Symp. Submar. Mass Mov. Conseq.* **2010**, *28*, 117–125.
11. Schofield, A.N. Use of centrifugal model testing to assess slope stability. *Rev. Can. Géotechnique* **2011**, *15*, 14–31. [CrossRef]
12. Chen, W.; Pourghasemi, H.R.; Kornejady, A.; Zhang, N. Landslide spatial modeling: Introducing new ensembles of ANN, MaxEnt, and SVM machine learning techniques. *Geoderma* **2017**, *305*, 314–327. [CrossRef]

13. Marjanović, M.; Bajat, B.; Abolmasov, B.; Kovačević, M. Machine Learning and Landslide Assessment in a GIS Environment. In *GeoComputational Analysis and Modeling of Regional Systems*; Springer: Berlin/Heidelberg, Germany, 2018.
14. Pham, B.T.; Prakash, I.; Bui, D.T. Spatial prediction of landslides using a hybrid machine learning approach based on Random Subspace and Classification and Regression Trees. *Geomorphology* **2018**, *303*, 256–270. [CrossRef]
15. Tse, K.C.; Chiu, H.; Tsang, M.; Li, Y.; Lam, E.Y. An unsupervised learning approach to study synchroneity of past events in the South China Sea. *Front. Earth Sci.* **2019**, *13*, 628–640. [CrossRef]
16. Qi, C.; Tang, X. Slope stability prediction using integrated metaheuristic and machine learning approaches: A comparative study. *Comput. Ind. Eng.* **2018**, *118*, 112–122. [CrossRef]
17. Ghorbanzadeh, O.; Blaschke, T.; Gholamnia, K.; Meena, S.; Tiede, D.; Aryal, J. Evaluation of Different Machine Learning Methods and Deep-Learning Convolutional Neural Networks for Landslide Detection. *Remote Sens.* **2019**, *11*, 196. [CrossRef]
18. Hong, H.; Liu, J.; Zhu, A.X. A novel hybrid integration model using support vector machines and random subspace for weather-triggered landslide susceptibility assessment in the Wuning area (China). *Environ. Earth Sci.* **2017**, *76*, 652. [CrossRef]
19. Yang, Z.S.; Chen, W.; Chen, Z. Subaqueous Landslides system in the Huanghe River (Yellow River) Delta. *Oceanol. Limnol. Sinica* **1994**, *20*, 573–581. Available online: http://en.cnki.com.cn/Article_en/CJFDTOTAL-HYFZ199406000.htm (accessed on 31 August 2022).
20. Sun, Y.F.; Hu, G.H.; Song, Y.P. *Study on the Key Technology of Prediction, Evaluation and Prevention of Offshore Submarine Geological Hazards*; First Institute of Oceanography, MNR: Qingdao, China, 2016.
21. Mahmood, K.; Kim, J.M.; Ashraf, M.; Ziaurrehman. The Effect of Soil Type on Matric Suction and Stability of Unsaturated Slope under Uniform Rainfall. *KSCE J. Civ. Eng.* **2016**, *20*, 1294–1299. [CrossRef]
22. Łukasik, S.; Kowalski, P.A.; Charytanowicz, M.; Kulczycki, P. Clustering using flower pollination algorithm and Calinski-Harabasz index. In Proceedings of the IEEE Congress on Evolutionary Computation (CEC), Vancouver, BC, Canada, 24–29 July 2016; pp. 2724–2728. [CrossRef]
23. Thangavel, K.; Aranganayagi, S. Clustering Categorical Data Using Silhouette Coefficient as a Relocating Measure. In Proceedings of the International Conference on Computational Intelligence and Multimedia Applications (ICCIMA 2007), Sivakasi, India, 13–15 December 2007; pp. 13–17. [CrossRef]
24. Davies, D.L.; Bouldin, D.W. A Cluster Separation Measure. *IEEE Trans. Pattern Anal. Mach. Intell.* **1979**, *1*, 224–227. [CrossRef]
25. Macqueen, J. Some Methods for Classification and Analysis of Multi Variate Observations. *Proc Berkeley Symp. Math. Stat. Probab.* **1965**, *5*, 281–297. Available online: https://www.docin.com/p-542657058.html (accessed on 31 August 2022).
26. Pedregosa, F. Scikit-learn: Machine Learning in Python. *JMLR* **2011**, *12*, 2825–2830.
27. Xiao, P.; Li, A.L. Prediction of Submarine Landslide Stability Based on GIS in the Yellow River Subaqueous Delta. *Geol. Sci. Technol. Inf.* **2016**, *35*, 221–226.
28. Sun, Y.F.; Huang, B.L. A Potential Tsunami impact assessment of submarine landslide at Baiyun Depression in Northern South China Sea. *Geoenviron. Disasters* **2014**, *1*, 1–11. [CrossRef]
29. Masson, D.G.; Harbitz, C.B.; Wynn, R.B.; Pedersen, G.; Løvholt, F. Submarine landslides: Processes, triggers and hazard prediction. *Philos. Trans. R. Soc. A Math. Phys. Eng. Sci.* **2006**, *364*, 2009–2039. [CrossRef]
30. James, J.J. Development of a Database and Assessment of Seafloor Slope Stability Based on Published Literature. Ph.D. Thesis, University of Texas at Austin, Austin, TX, USA, 2003.
31. Gan, H.Y.; Wang, J.S.; Hu, G.W. Submarine Landslide Related to Natural Gas Hydrate within Benthal Deposit. *J. Seismol.* **2004**, *24*, 177–181. [CrossRef]
32. Jiang, M.J.; Sun, C.; Crosta, G.B.; Zhang, W.C. A study of submarine steep slope failures triggered by thermal dissociation of methane hydrates using a coupled CFD-DEM approach. *Eng. Geol.* **2015**, *190*, 1–16. [CrossRef]

applied sciences

MDPI

Article

Automatic Identification of Landslides Based on Deep Learning

Shuang Yang [1], Yuzhu Wang [1,*], Panzhe Wang [2], Jingqin Mu [1,3], Shoutao Jiao [4], Xupeng Zhao [1], Zhenhua Wang [1], Kaijian Wang [5] and Yueqin Zhu [6]

1 School of Information Engineering, China University of Geosciences, Beijing 100083, China
2 School of Geophysics and Information Technology, China University of Geosciences, Beijing 100083, China
3 Department of Computer Science, Tangshan Normal University, Tangshan 063000, China
4 Development and Research Center, China Geological Survey, Beijing 100037, China
5 China Aero Geophysical Survey and Remote Sensing Center for Natural Resources, China Geological Survey, Beijing 100083, China
6 National Institute of Natural Hazards, Ministry of Emergency Management, Beijing 100085, China
* Correspondence: wangyz@cugb.edu.cn

Abstract: A landslide is a kind of geological disaster with high frequency, great destructiveness, and wide distribution today. The occurrence of landslide disasters bring huge losses of life and property. In disaster relief operations, timely and reliable intervention measures are very important to prevent the recurrence of landslides or secondary disasters. However, traditional landslide identification methods are mainly based on visual interpretation and on-site investigation, which are time-consuming and inefficient. They cannot meet the time requirements in disaster relief operations. Therefore, to solve this problem, developing an automatic identification method for landslides is very important. This paper proposes such a method. We combined deep learning with landslide extraction from remote sensing images, used a semantic segmentation model to complete the automatic identification process of landslides and used the evaluation indicators in the semantic segmentation task (mean IoU [mIoU], recall, and precision) to measure the performance of the model. We selected three classic semantic segmentation models (U-Net, DeepLabv3+, PSPNet), tried to use different backbone networks for them and finally arrived at the most suitable model for landslide recognition. According to the experimental results, the best recognition accuracy of PSPNet is with the classification network ResNet50 as the backbone network. The mIoU is 91.18%, which represents high accuracy; Through this experiment, we demonstrated the feasibility and effectiveness of deep learning methods in landslide identification.

Keywords: deep learning; semantic segmentation; PSPNet; landslide

Citation: Yang, S.; Wang, Y.; Wang, P.; Mu, J.; Jiao, S.; Zhao, X.; Wang, Z.; Wang, K.; Zhu, Y. Automatic Identification of Landslides Based on Deep Learning. *Appl. Sci.* **2022**, *12*, 8153. https://doi.org/10.3390/app12168153

Academic Editor: Rubén Usamentiaga

Received: 13 July 2022
Accepted: 11 August 2022
Published: 15 August 2022

1. Introduction

Landslides are common and frequent geological hazards around the world. The occurrence of landslides affects the terrain and causes different degrees of damage [1]. Furthermore, when residential areas or public buildings are close to a landslide site, the event is often accompanied by emergencies. Landslides in specific areas need to be identified within a short period to intervene and resolve the crisis [2,3]. With the rapid development of remote sensing technology, these methods have been widely used [4,5]. Currently, the landslide identification methods based on remote sensing images are mainly divided into visual interpretation, pixel-based and object-oriented landslide identification methods.

Visual interpretation is the earliest landslide identification method applied to remote sensing images. Visual interpretation is when the interpreter extracts the landslide shape from the remote sensing image according to his or her professional knowledge and related research materials. Data extracted by this method have high accuracy. However, the visual interpretation also has the disadvantages of being time-consuming, having an overrelian on manual discrimination and inefficient [6].

Pixel-based landslide identification methods focus on pixel values and pixel value changes in remote sensing images and classify them according to pixel changes or change characteristics [7]. Although this method overcomes the shortcomings of visual interpretation, it is only judged by a single pixel value; moreover, it does not consider the correlation between pixels, resulting in an indistinct recognition of landslide edge regions and poor performance [8].

Object-oriented landslide identification methods utilize the attribute features of raw data (such as texture, spectrum, etc.) to classify remote sensing images by one or more attributes [9]. However, this method uses one or more attribute features to set thresholds, and the process is complicated and cannot process large-scale remote sensing images in time.

With the continuous increase in high-resolution remote sensing images, how to quickly and efficiently identify targets from massive remote sensing images has become a key problem for scholars and experts to study. In recent years, with the rapid development of deep learning, computer vision and image processing technology have been introduced into the field of remote sensing as a new method for remote sensing image classification and target detection [10]. Research has shown that deep learning methods do not require many of the data provided in traditional landslide identification methods to help identify landslides; they only need enough landslide images as samples and training, which greatly simplifies the complex calculation process, makes up for the shortcomings of the above-mentioned traditional landslide identification methods, and realizes automatic identification of landslides. At the same time, deep learning methods are higher accuracy than the traditional landslide identification methods [11]. The main contributions of this article are as follows:

(1) We process the landslide data in the Bijie landslide dataset, create a landslide dataset, and preprocess the dataset (data cleaning, data enhancement);
(2) On the landslide dataset, we use three models (U-Net, DeepLab v3+ and PSPNet) to conduct experiments and test the performance changes in the models when different classification networks are used as the backbone network;
(3) We use the above pretrained model to test the landslide test set and use mIoU, precision, and recall to evaluate the model performance to obtain the optimal model for landslide identification performance.

The remainder of this paper is structured as follows. Section 2 introduces related work. Section 3 shows the data and methodology we used. Section 4 presents and analyzes the experiment results. Section 5 presents our conclusions.

2. Related Work

Convolutional neural networks (CNN) [12] have achieved great success in the field of image processing because of their nonlinear learning ability [13], driving the rapid development of computer vision [14,15]. Based on CNN studies, various models have been developed for image classification [16], object detection [17,18], semantic segmentation [19], etc., where semantic segmentation performs pixel-level segmentation of the images. These models have achieved satisfactory results in traditional vision tasks. Therefore, people have begun to apply these deep learning models to landslide identification in remote sensing images in the past few years.

Ye et al. (2019) proposed a constrained deep learning model, applied it to identifying landslides in hyperspectral images, and compared the results with the support vector machine–spectral information divergence–spectral angle matching method. They that the extraction of high-level features by deep learning has great potential for improving the accuracy of landslide identification [20]. Ghorbanzadeh et al. (2019) used CNNs for Himalayan landslide identification and compared them with state-of-the-art machine learning methods (artificial neural networks, support vector machines, and random forests); the results show that deep learning is superior to machine learning in landslide identification experiments [21]. Prakash et al. (2020) proposed an improved U-Net model that uses ResNet34 blocks for feature extraction and enables landslide identification in Douglas County, south of Portland, Oregon, USA [22]. Zhu et al. (2020) proposed a method based on U-Net architecture to fuse local and nonlocal features, upsampling by dilated convolution, and the corresponding spatial pyramid expanded receptive field and scale attention mechanism to identify the landslide caused by the earthquake in Jiuzhaigou, China [23]. Ji et al. (2020) developed a CNN-based spatial channel attention mechanism to classify and identify landslides in Bijie City, China, from available satellite imagery and DEM datasets; this experiment concludes that the attention mechanism and DEM data can effectively improve the accuracy of landslide identification [24]. Liu (2020) proposed to use ResU-Net to identify earthquake landslides in Jiuzhaigou, Sichuan Province, China, and obtained an F1 value of 93.3%, and an mIoU value of 87.5% [25]. Ju et al. (2020) identified old loess landslides on Google Earth images using the two-stage algorithm Mask R-CNN; although the accuracy rate did not reach a high level, it confirmed the feasibility of Mask R-CNN to identify old landslides [26]. Dai et al. (2021) proposed an improved U-Net neural network and completed the automatic identification of the deformation features of the landslide time series [27]. Ullo et al. (2021) used the Mask R-CNN model with ResNet101 as the backbone network and the transfer learning algorithm to complete landslide recognition in digital images of hilly areas obtained by drones, and the results show that the method is superior to the existing research in both algorithm performance and robustness [28]. Liu (2021) proposed to use an improved Mask R-CNN to identify earthquake landslides in the Jiuzhaigou area of Sichuan Province, China, and obtained an F1 value of 94.5% and an mIoU value of 89.6% [29]. Ghorbanzadeh (2022) combined the object-based image analysis (OBIA) approach with the fully convolutional network (FCN) model to complete the landslide detection in Sentinel-2 images and verified the method's feasibility [30].

Therefore, deep learning methods have been applied to landslide identification. However, due to the diversity and complexity of landslides, these methods still have many problems to be solved. For example, in order to improve the ability to identify landslides, the model needs to learn a large number of data [31]; this is a key issue because there is very little landslide data currently available. In addition, for this work, more data information can help the model to better improve the landslide recognition accuracy [32]; however, the acquisition, retrieval, and annotation of datasets is often a difficult point in landslide identification tasks. Therefore, to address the issues mentioned above, we conducted this study. Since there are few publicly available landslide datasets and their quality is uneven, to have a good experimental basis, we selected the Bijie dataset published by Ji et al. (2020) [24]. The Bijie dataset is the first large-scale, public remote sensing landslide dataset and has a double check; more detailed dataset information will be introduced in Section 3.1. At the same time, since the interpretation of landslide images has very high professional requirements, in the re-labeling of samples, we strictly follow the samples provided by the Bijie dataset to ensure the reliability of the data. In the final sample set, we expand the sample set through the data augmentation method and finally obtain a data set containing 2500 landslide images.

3. Materials and Methods

3.1. Data Source

Remote sensing datasets of landslide are difficult to obtain, we used the open source Bijie landslide dataset [24]. The Bijie landslide dataset is the first open remote sensing landslide dataset with careful triple inspection; the data set was proposed by scholars such as Ji et al. (2020), and the classification research of landslides was carried out on it. Its study area is located in Bijie City, Guizhou Province, China, with about 26,853 square km and an altitude ranging from 457 m to 2900 m. The soil on the slopes caused by the perennial rainfall is soft and prone to landslides, and it is one of the most prone areas in China.

The remote sensing images in the Bijie landslide dataset were captured by the TripleSat satellite, and the RGB images have a resolution of 0.8 m. Seven hundred and seventy landslide images and two thousand and three other types of images were intercepted from the captured remote sensing images. The dataset consists of satellite optical images and label files. In the process of making the dataset, two methods were adopted to interpret the landslide images to ensure the reliability of the database: One is the visual interpretation by geologists through optical remote sensing images; the other is based on residents' reports and field surveys. Throughout the work, the shapes of landslide samples were drawn with the help of ArcGIS.

3.2. U-Net

U-Net is a semantic segmentation network proposed by Olaf Ronneberger in the ISBI Cell Segmentation Competition in 2015. It utilizes a U-shaped network structure to capture contextual information and location information. It was initially used to solve medical image segmentation problems, especially cell-level segmentation tasks, and was gradually used to solve problems in other fields [33].

The network structure of U-Net, which is an encoder-decoder structure, is shown in Figure 1.The encoder utilizes the idea of stacking convolutional layers, downsampling the feature map through convolution and pooling, and performing four total pooling operations. After each stacking convolution layer operation, the size of the feature map is halved and, at the same time, the pooling result of each step is passed to the decoder; in the decoder, the feature map is first upsampled or deconvolved and then concatted on the channel with the previous feature map of the same size. Convolution and upsampling is then performed, and after upsampling four times, an output result of the same size as the original image is obtained.

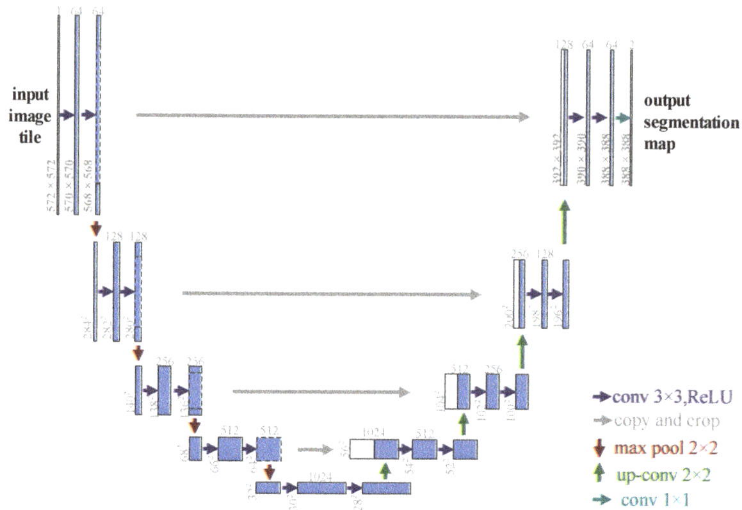

Figure 1. U-Net architecture. The blue box represents the multichannel feature layer. The channel number is shown at the top of the box. The white boxes represent the replicated feature maps. The arrows represent operations on the feature layers.

3.3. DeepLab v3+

DeepLabv3+ was proposed by the Google team in 2018 and is the DeepLab series model [34]. DeepLabv3+ is based on DeepLabv3 and improves it. It uses Deeplabv3 as the encoder, introduces atrous convolution in the encoder for downsampling and uses the spatial pyramid pooling module to extract multiscale information, which improves the accuracy by fusing low-level and high-level features.

Its specific structure is shown in Figure 2. The encoder extracts image features through a deep convolutional neural network (DCNN), and the extracted feature layers are input to the decoder for 1×1 convolution. Meanwhile, the feature layers extracted by DCNN use $1 \times 1, 3 \times 3, 3 \times 3$, and 3×3 atrous convolutions for downsampling and pooling, where the expansion rates of atrous convolution are 1, 6, 12 and 18, respectively. Then, the obtained new feature layer is concatted, its channel is changed to 1/5 of the original through 1 × 1 convolution, and the decoder is entered for upsampling. The unsampled feature layer is concatted with the feature layer in the decoder, and then the 3 × 3 convolution and upsampling are gone through to obtain the final prediction map.

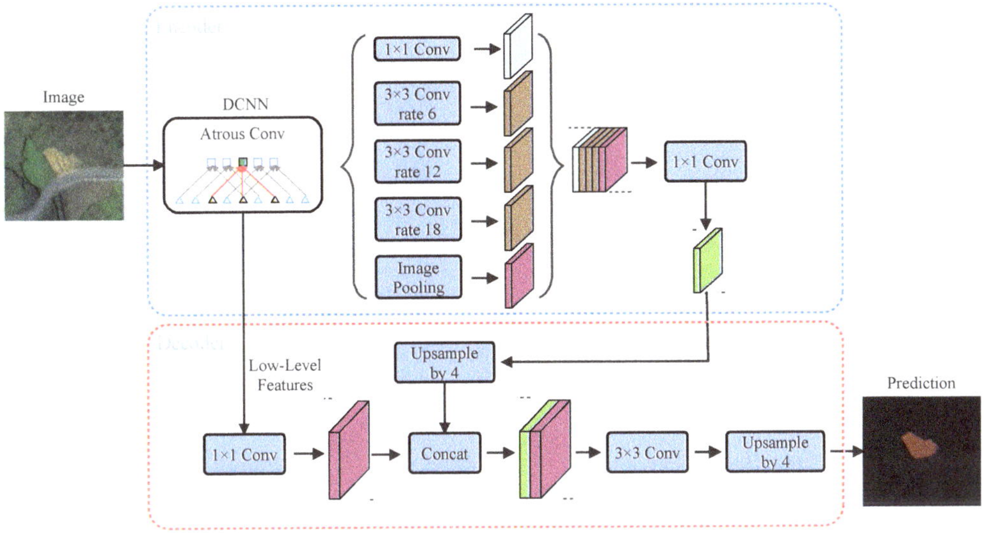

Figure 2. DeepLabv3+ architecture. DCNN stands for deep convolutional neural network, and Atrous Conv stands for atrous convolution. Cyan, orange, pink and green represent the extracted feature layers. The dark blue boxes represent the operations taken on the feature layers.

3.4. PSPNet

PSPNet is a semantic segmentation model jointly proposed by the Chinese University of Hong Kong and Shangtang Technology and it won the championship in the 2016 ImageNet Challenge [35]. The original intention of PSPNet was to improve the FCN. The most prominent feature of PSPNet is that it adds a PSP module between the encoder and the decoder, which is also the main difference between it and the FCN.

The structure of PSPNet is shown in Figure 3; the input layer obtains the feature layer of the input image through CNN, and the feature layer size is changed to 1/5 of the original through. Then, the obtained feature map is input to the pyramid pooling module. First, this module divides the input feature layer into 6 × 6, 3 × 3, 2 × 2 and 1 × 1 sized areas; the ave-pooling operation is performed in the divided area to obtain four feature layers of different sizes (corresponding to the green, blue, orange and red outputs in Figure 3, respectively); then, 1 × 1 convolution operations are performed on these feature layers; next, the number of channels of the feature layer is changed to one-fourth of the original; and finally, the feature layer is up-sampled by bilinear interpolation. The upsampled feature map and the feature layer obtained by CNN are concatenated, and finally, the final output is obtained through the convolution operation.

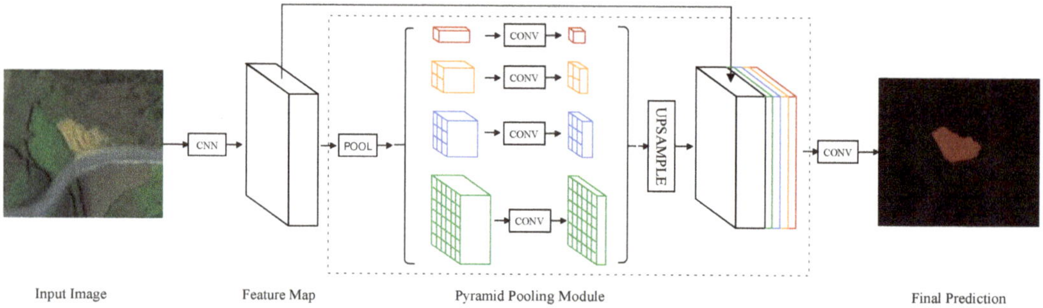

Figure 3. PSPNet architecture. First, a CNN is used to obtain the last convolutional feature map given an input image. A pyramid parsing module is then applied to collect different subregion representations, followed by upsampling and concatenation layers to form the final feature representation. Finally, the representation is fed into convolutional layers to obtain the final per-pixel predictions.

3.5. Evaluation Metrics

In deep learning methods, recall rate and accuracy rate are indicators that can evaluate the recognition effect of the model, and they are associated with the confusion matrix; As Table 1 shows, T and F represent the prediction of true or false; P (positive) and N (negative) represent the type of prediction; TP (true positive), TN (true negative), FP (false positive), and FN (false negative) are used to classify pixels. Taking this article as an example, TP means that the pixel is identified as a landslide pixel; the identification is correct. TN means that the pixel is identified as a nonlandslide pixel; the identification is correct. FP indicates that the pixel was identified as a landslide pixel and identified incorrectly. FN indicates that the pixel was identified as a nonlandslide pixel and identified incorrectly.

Table 1. Confusion matrix of classification results.

Actual Values	Predicted Values	
	Positive	**Negative**
Positive	TP	FN
Negative	FP	TN

Precision represents the proportion of the actual landslide pixels in the pixels predicted by the model as landslides, as shown in Equation (1):

$$Precision = \frac{TP}{FP + TP}, \quad (1)$$

Recall represents the proportion of landslide pixels predicted by the model in all actual landslide pixels, as shown in Equation (2):

$$Recall = \frac{TP}{FN + TP}, \quad (2)$$

In addition, mIoU is a widely used metric in semantic segmentation tasks and is used as a standard measure to measure semantic segmentation models. Intersection over union (IoU) represents the ratio between the intersection and union of the predicted results of landslide pixels and the actual landslide pixels, and mIoU represents the average of all categories of IoU. The IoU is shown by Equation (3):

$$mIoU = \frac{1}{k+1} \sum_{i=0}^{k} \frac{p_{ii}}{\sum_{j=0}^{k} p_{ij} + \sum_{j=0}^{k} p_{ji} - p_{ii}}, \quad (3)$$

In this formula, i represents the real pixel, j represents the predicted pixel, p_{ij} represents that i will be predicted as j, p_{ii} represents that i will be predicted as i and p_{ji} represents that j will be predicted as i.

4. Results and Discussion

4.1. Data Preprocessing

Data preprocessing includes relabeling samples and implementing data augmentation strategies on samples. Due to different target tasks, we only selected 770 landslide samples in the Bijie landslide dataset as the original data. In the labeling work, in order to obtain reliable data, we relabeled according to the original labels provided in the Bijie dataset. At the same time, we processed the data according to the research goals and finally obtained 510 landslide samples. Among these obtained data, 90% of the data was used for training and validation of the model, and 10% of the data was used for testing the model.

The deep learning model learns the features of the image through the provided image samples during the training process. Therefore, the more samples that are provided to the model, the more the model can learn the features of such images and the better its predictions. For the model, if an image is rotated, cropped and then passed into the model, the model will consider it a new image, so we can enhance the sample by expanding it. We implemented a data enhancement strategy for the existing training set according to the characteristics of sample landslides with different directions, different structures and different boundary shapes. We enhance the dataset's quality through augmentation techniques, we thereby improved the model's training effect. However, excessive rotation and flipping of images will cause overfitting of the model, thereby reducing its generalization and causing causing the model to achieve high accuracy on the training set but but not achieve very high accuracy on the test set. Considering these problems in the process of data expansion, we rotate, scale and flip the sample set according to a certain probability based on experience. These expanded data were all used in our model training; after expansion, the training samples were changed from the original 510 to 2500. For more details on data preprocessing, see Table 2.

Table 2. Data augmentation.

Method	Probability of Execution	Specific Operations
random rotation	50%	rotate 20°, +90°, −90°
left-right flipping	100%	flip the image left and right
image cropping	100%	original image 0.7× dimension

4.2. Training

The training effect of the deep learning model is closely related to the accuracy of the data set, suitable parameters, and training methods. Therefore, in this study, we selected three models, U-Net, DeepLab v3+, and PSPNet, and used two different classification networks as the backbone network of each model; the backbone network selection of the model is shown in Table 3. We record these models as U-Net (VGG), U-Net (ResNet50), DeepLab v3+ (MobileNet), DeepLab v3+ (Xception), PSPNet (MobileNet), and PSPNet (ResNet50).

In the experiment, we set and adjusted the training parameters uniformly for all the models, as shown in Table 4. The input image size is fixed at 473 × 473. The classification of pixel types is landslide and background. The model training adopts the method of freezing training, which divides the training into two stages: freezing and unfreezing. In the freezing stage, the backbone of the model is frozen, and the feature extraction network does not change. At this time, 50 rounds of fine-tuning are performed on the network. The video memory occupied in the freezing phase is small, so the batch_size and learning rate are set larger. In the unfreezing stage, the backbone network of the model is unfrozen, and the feature extraction network will change; at this time, the network is trained for

100 rounds. Due to a large amount of video memory occupied, the batch_size is set to 0.5 times that of the frozen phase, and the learning rate is reduced.

Table 3. The backbone network selection of the models.

Model	Backbone
U-Net	VGG
U-Net	ResNet50
DeepLab v3+	MobileNet
DeepLab v3+	Xception
PSPNet	MobileNet
PSPNet	ResNet50

Table 4. Parameters for our models.

Hyper-Parameter	Parameter Values
input_shape	[473, 473]
classes	landslide, background
freeze_Train	True
pretrained weights	True
datasets used for pre-training	VOC data set
Init_Epoch	0
downsample_factor	8
freeze_epoch	50
unfreeze_epoch	100
freeze_learning_rate	10^{-4}
freeze_batch_size	8
unfreeze_batch_size	4
unfreeze_learning_rate	10^{-5}
focal_loss	True
dice_loss	True
eager pattern	False
aux_branch	False
early_stopping	True
num_workers	1
cls_weights	np.array([1, 2], np.float32)

In addition, during model training, we use the pre-trained weights obtained in the VOC dataset as the initial parameters of the model and set dice_loss to balance the number of training categories. We set focal_loss to balance positive and negative samples and use the NumPy form to give different loss weights to the background and landslides so that the model focuses on landslide pixels. In order to improve the recognition effect, we set the downsample_factor to 8, but it will also occupy much memory; therefore, in order to reduce the occupation of video memory, we do not use aux_branch by default, do not use multi-threading to read data and use the early stop strategy to save computing resources.

The hardware environment of this experiment: GPU: 4*NVIDIA Tesla K80, CPU: 32*Intel (R) Xeon (R) CPU E5-2620 v4 @ 2.10GHz, OS: CentOS 8.3. The software environment of this experiment: CUDA 11.2, Python 3.6, PyTorch 1.10.1, Tensorflow 2.2.0.

4.3. Experimental Results and Analysis

Pretrained models obtained by the experiment were used to predict the test set. The prediction results show that these models can effectively identify landslides. At the same time, according to the experimental results, we conclude that the PSPNet model using ResNet50 as the backbone network has the best recognition effect. We will discuss the results first from the perspective of image recognition and then from the perspective of index evaluation.

A total of 51 remote sensing images of landslides were included in the test set; all of the images were from the eastern part of the Qinghai–Tibet Plateau in Bijie City, Guizhou Province, China. As shown in Figure 4, we show some landslide images in the test set, and none of the images in the test set participated in the training, through which we evaluated the performance of each model. We have selected some of these samples for analysis, as shown in Figure 5. The figure includes four samples of landslide images that occurred in different places, marked as Landslide I, Landslide II, Landslide III, and Landslide IV. Among them, I, II, and III are new landslides with various characteristics and shapes of landslides, and IV is an old landslide with inconspicuous characteristics of landslides. In addition, the figure also includes the label file of the landslide, which was used for comparison with the predicted map. We use the pretrained model to obtain the predicted map; we will analyze the predicted maps of the four landslide samples separately.

Figure 4. Examples of landslide samples in the test set, all from Bijie, Guizhou, China.

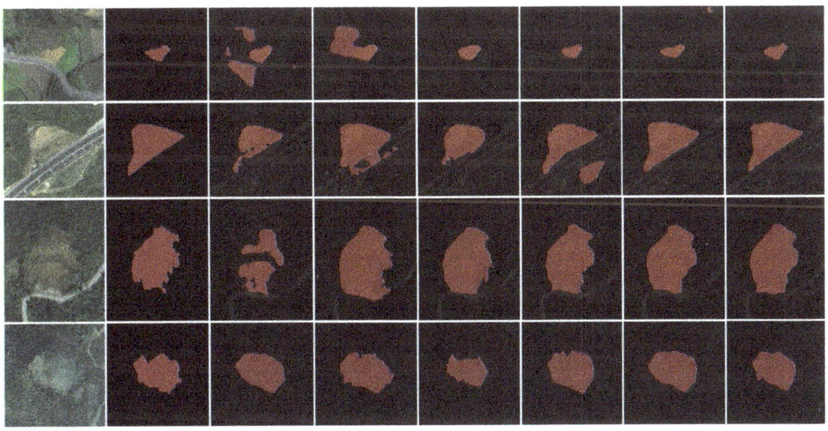

Figure 5. Model prediction of landslide results. Horizontally, they are Landslide I, Landslide II, Landslide III and Landslide IV. Vertically, they are the landslide image, label map (for comparison) , U-Net (VGG) recognition map, DeepLabv3+ (Xception) recognition map, DeepLab v3+ (MobileNet) recognition map, U-Net (ResNet50), the recognition map of PSPNet (MobileNet) and the recognition map of PSPNet (ResNet50).

Landslide I: U-Net (VGG) has a high error rate in identifying landslide images; it identifies the background associated with the landslide color as a landslide. However, the majority of these environments are land and cultivated land. There are 15 additional prediction maps for the same situation. DeepLab v3+ (Xception) is similar to U-Net (VGG). DeepLab v3+ (MobileNet), U-Net (ResNet50), PSPNet (MobileNet), and PSPNet (ResNet50) can more accurately identify landslides; in contrast, PSPNet (ResNet50) has better performance in recognizing landslides.

Landslide II: U-Net (VGG) and DeepLab v3+ (MobileNet) cannot fully identify landslides, so the effect is poor. DeepLabv3+ (Xception) recognizes roads with similar colors as landslides. U-Net (ResNet50) recognizes other objects with similar shapes to landslides as landslides and cannot distinguish the landslide from other content in the background. PSPNet (MobileNet) and PSPNet (ResNet50) perform better.

Landslide III: There is a chasm in identifying the landslide by U-Net (VGG); the color of the chasm part is darker, so the background color is closer, which is caused by the new vegetation growing on the landslide, the model cannot recognize this and thus identifies errors. The situation of DeepLab v3+ (Xception) is just the opposite. Although it can distinguish landslides from vegetation, it cannot distinguish the features between landslides and roads. DeepLabv3+ (MobileNet), U-Net (ResNet50), PSPNet (MobileNet) and PSPNet (ResNet50) can identify landslides more accurately; however, their sensitivities to landslide boundaries vary.

Landslide IV: Landslide IV has been formed for a long time, the landslide has been covered with vegetation, and the entire landslide is green. Because its landslide characteristics are not prominent, they are not easy to separate. This type of landslide is too complex for the model to recognize. Fortunately, according to the recognition effect of these models on Landslide IV, the model can distinguish the landslide from the features of the surrounding vegetation. However, the segmentation effect of the boundary is not accurate enough.

We evaluated the model using the metrics in Section 3.5, and the evaluation results are shown in Table 5. Observing the recall index and precision index values, we found that the recall rate of all models is higher than the precision; this means that these models can identify real landslide pixels but also identify many non-landslide pixels as landslide pixels. It can also be seen from the analysis of Figure 6 that when identifying Landslide I and Landslide II, U-Net (VGG), DeepLabv3+ (Xception) and U-Net (ResNet50) easily confuse other objects with similar colors and shapes to landslides.

In Table 5, mIoU values are used to evaluate model performance comprehensively. Among these pretrained models, PSPNet (ResNet50) produced the best landslide recognition effect, with an mIoU value of 91.18%, and obtained the highest precision index (93.76%); this means that the model has a good effect on the recognition of landslide pixels. Followed by PSPNet (MobileNet) and U-Net (ResNet50), the mIoU values of which are 89.11% and 88.75%, respectively; PSPNet (MobileNet) obtained the highest recall index (97.39%), which means that the model can identify most of the landslide pixels. U-Net (VGG) has the worst landslide recognition effect, with a mIoU of 81.64% and a recall and precision of 89.34% and 89.22%, respectively.

Table 5. The results of the suggested model. Numbers in bold represent the best model for identifying landslides (with mIoU metric as the final criterion).

Model	Backbone	mIoU	Recall	Precision
U-Net	VGG	81.64%	89.34%	89.22%
DeepLab v3+	Xception	86.15%	92.26%	92.20%
DeepLab v3+	MobileNet	87.06%	94.06%	91.64%
U-Net	ResNet50	88.75%	96.15%	91.82%
PSPNet	MobileNet	89.11%	97.39%	92.61%
PSPNet	**ResNet50**	**91.18%**	**96.9%**	**93.76%**

Below, we combine the chart to discuss the recognition effect of PSPNet on landslides when MobileNet and ResNet50 are used as the backbone network, respectively. As shown in Table 6, we used precision, recall and IoU to evaluate the model's ability to recognize landslide and background pixels, respectively, when using two different backbone networks. P is the abbreviation for precision, R is the abbreviation for recall and IoU represents intersection over union. In identifying background pixels, when ResNet50 is used as the backbone network, the IoU value is 97.76%, which is 14.79% higher than when MobileNet is used as the backbone network; this can be seen intuitively from the landslide prediction map. In Figure 6, the blue boxes mark the parts of PSPNet (ResNet50) and PSPNet (MobileNet) that misidentify the background pixels as landslide pixels. Compared with PSPNet (MobileNet), PSPNet (ResNet50) is less likely to mistakenly identify content in the background (such as roads, green vegetation, bare land) as landslides. At the same time, the precision and recall of ResNet50 are higher than MobileNet. In identifying landslide pixels, when ResNet50 is used as the backbone network, the IoU value is 84.6%, which is 3.45% higher than when MobileNet is used as the backbone network. At the same time, the precision is 5.35% higher than that of PSPNet (MobileNet), and the recall is lower than 1.9%. Therefore, its landslide recognition effect is better.

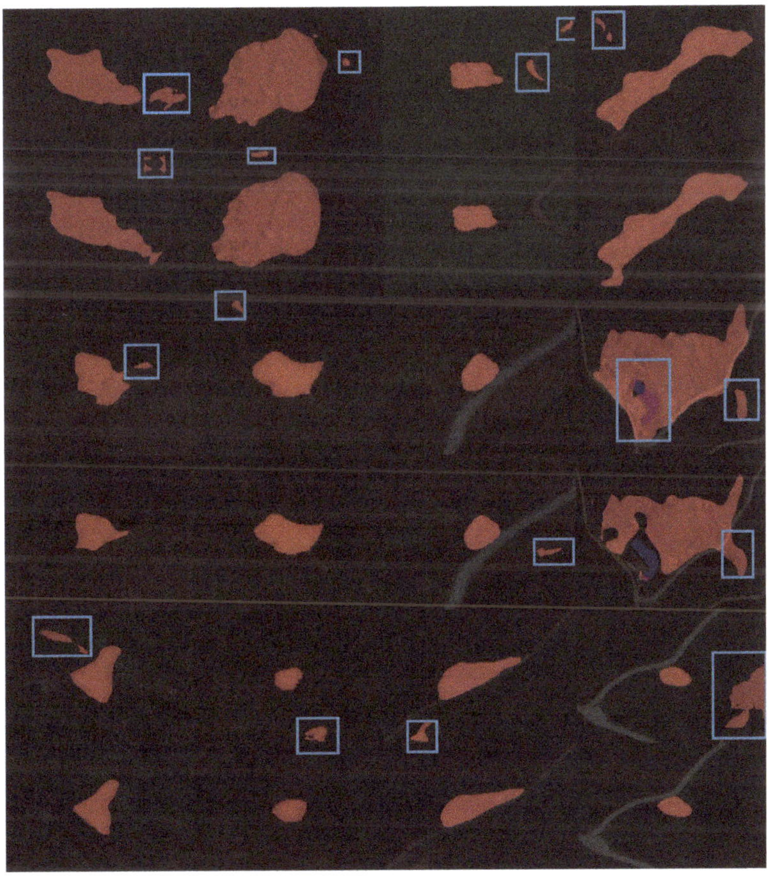

Figure 6. The comparison results of landslide identification of PSPNet using MobileNet and ResNet50 as the backbone network, respectively. Behaviors 1, 3 and 5 use MobileNet as the identification result of the backbone network, and behaviors 2, 4 and 6 use ResNet50 as the identification result of the backbone network. The blue box marks the background pixels that the model misidentified as part of the landslide.

Table 6. The performance of PSPNet under different backbone networks. The numbers in bold are the highest IoU values obtained for landslide and background recognition, respectively.

Backbone	Evaluate	Landslide	Background
MobileNet	P	82.79%	97.08%
	R	97.37%	97.41%
	IoU	81.15%	82.97%
ResNet50	P	88.14%	99.41%
	R	95.47%	98.34%
	IoU	84.6%	97.76%

4.4. Discussion

In previous research on landslide recognition based on deep learning, Ghorbanzadeh used CNN to train and test landslides in the Himalayas and obtained an F1 value of 87.8% and an mIoU value of 78.26% [21]. Liu (2020) proposed to use ResU-Net to identify earthquake landslides in Jiuzhaigou, Sichuan Province, China, and obtained an F1 value of 93.3% and an mIoU value of 87.5% [25]. Ullo (2021) applied Mask R-CNN to landslide recognition in digital images of target hilly areas acquired by drones; and when ResNet101 was used as the backbone network, the obtained F1 value was 97% [28]. Liu (2021) proposed to use an improved Mask R-CNN to identify earthquake landslides in the Jiuzhaigou area of Sichuan Province, China, and obtained an F1 value of 94.5% and an mIoU value of 89.6% [29]. Ghorbanzadeh (2022) obtained an F1 of 84.03% and an mIoU value of 72.49% when using ResU-Net and OBIA for landslide detection in multitemporal Sentinel-2 images [21]. Our proposed PSPNet, using the classification network ResNet50 as the backbone network, achieves a 91.18% mIoU value on the Bijie landslide dataset; however, due to the differences in datasets and evaluation metrics, we cannot compare it with other models; however, according to the current experimental results, the method proposed in this paper is effective for landslide recognition.

Although PSPNet (ResNet50) achieves good results in landslide identification, it still has some shortcomings. For example, its segmentation of landslide boundaries still needs further improvement. Landslide images are different from traditional remote sensing images. In addition to the information of the images themselves, remote sensing images also contain rich geological information. For example, a digital elevation model (DEM) can reflect local terrain features at a specific resolution. Therefore, we should further combine deep learning with remote sensing to maximize the role of remote sensing data. If the DEM data and remote sensing images are fused, we can obtain the local terrain information of the landslide from the DEM, which will help the model to improve the segmentation accuracy of the landslide boundary.

At present, the automatic identification of landslides based on deep learning still presents research challenges and problems to be solved. For example, the scarcity of open source code for landslide identification research and the lack of high-resolution public landslide remote sensing image datasets and validation areas have brought great difficulties to such research. At the same time, the further improvement in landslide identification accuracy remains to be explored. Given these problems, we need to continue research on automatic landslide identification. In terms of datasets, we will try to integrate the DEM data into remote sensing images so that the datasets contain more information about landslides, thereby improving the accuracy of landslide identification. In terms of models, we will further discuss the influence of model structure on landslide identification and then improve the model to improve the effect of landslide identification.

5. Conclusions

In this study, we proposed a deep learning-based research method for the automatic identification of landslides and obtained good results on the Bijie landslide dataset. First, we reconstructed the dataset for semantic segmentation and preprocessed the landslide data based on the Bijie landslide dataset. We then used three models: U-Net, DeepLabv3+ and PSPNet, each using two classification networks as the backbone network, and completed training and validation on the Bijie landslide dataset on these models. Finally, we used the experimentally obtained pretrained model for landslide recognition and used mIoU, precision and recall to evaluate the model's recognition effect on landslides. According to the experimental results, we obtained the best recognition effect of the PSPNet model with ResNet50 as the backbone network, with an mIoU of 91.18%. This experimental result shows that it is feasible to detect the automatic identification of landslides by using the deep learning method; simultaneously, our proposed PSPNet method with ResNet50 as the backbone network can effectively identify landslides.

Based on the above experimental results, we believe that this research will be helpful to landslide relief operations in real life. The automatic identification method can effectively make up for the shortcomings methods, which are time-consuming, labor-intensive and highly dependent on labor; save much time and human resources for emergency rescue work; and reduce the loss of life and property. At the same time, it can help geologists significantly improve their work efficiency and allow them to spend more time on work that requires more geologists. Therefore, this research has significant practical potential.

In future work, we will contribute to the lack of high-resolution landslide remote sensing image datasets and open source code for landslide identification research. We will try to enrich the landslide data set by adding more factors so that it contains more landslide information for the model to improve the recognition accuracy of landslides further. In terms of models, we will try to explore, for example, the self-attention mechanism is introduced into the model so that the model can learn the characteristics of landslides in a more targeted way, thereby improving its accuracy. In addition, compared with new landslides, the characteristics of old landslides are less obvious, and it is not easy to realize automatic identification. Thus, we will also try to apply the training model of new landslides to the extraction of old landslides through transfer learning, thereby improving the accuracy of deep learning methods in the automatic identification of old landslides.

Author Contributions: Conceptualization, Y.W. and Y.Z.; methodology, S.Y. and Y.W.; software, S.Y., P.W. and X.Z.; validation, S.Y. and P.W.; writing—original draft preparation, S.Y.; writing—review and editing, Y.W., J.M., S.J., Z.W. and K.W.; supervision, Y.W. All authors have read and agreed to the published version of the manuscript.

Funding: This work was supported in part by the National Natural Science Foundation of China under Grant 41872253 and in part by the GHFUND B of China under Grant ghfund202107021958.

Institutional Review Board Statement: Not applicable.

Informed Consent Statement: Not applicable.

Data Availability Statement: Not applicable.

Conflicts of Interest: The authors declare no conflict of interest. The funders had no role in the design of the study; in the collection, analyses, or interpretation of data; in the writing of the manuscript, or in the decision to publish the results.

References

1. Hacıefendioğlu, K.; Demir, G.; Başağa, H.B. Landslide detection using visualization techniques for deep convolutional neural network models. *Nat. Hazards* **2021**, *109*, 329–350. [CrossRef]
2. Voigt, S.; Kemper, T.; Riedlinger, T.; Kiefl, R.; Scholte, K.; Mehl, H. Satellite image analysis for disaster and crisis-management support. *IEEE Trans. Geosci. Remote Sens.* **2007**, *45*, 1520–1528. [CrossRef]
3. Plank, S.; Twele, A.; Martinis, S. Landslide mapping in vegetated areas using change detection based on optical and polarimetric SAR data. *Remote Sens.* **2016**, *8*, 307. [CrossRef]
4. Czikhardt, R.; Papco, J.; Bakon, M.; Liscak, P.; Ondrejka, P.; Zlocha, M. Ground stability monitoring of undermined and landslide prone areas by means of sentinel-1 multi-temporal InSAR, case study from Slovakia. *Geosciences* **2017**, *7*, 87. [CrossRef]
5. Rosi, A.; Tofani, V.; Tanteri, L.; Tacconi Stefanelli, C.; Agostini, A.; Catani, F.; Casagli, N. The new landslide inventory of Tuscany (Italy) updated with PS-InSAR: geomorphological features and landslide distribution. *Landslides* **2018**, *15*, 5–19. [CrossRef]

6. Li, Z.; Shi, W.; Lu, P.; Yan, L.; Wang, Q.; Miao, Z. Landslide mapping from aerial photographs using change detection-based Markov random field. *Remote Sens. Environ.* **2016**, *187*, 76–90. [CrossRef]
7. Li, Z.; Shi, W.; Myint, S.W.; Lu, P.; Wang, Q. Semi-automated landslide inventory mapping from bitemporal aerial photographs using change detection and level set method. *Remote Sens. Environ.* **2016**, *175*, 215–230. [CrossRef]
8. Han, Y.; Wang, P.; Zheng, Y.; Yasir, M.; Xu, C.; Nazir, S.; Hossain, M.S.; Ullah, S.; Khan, S. Extraction of Landslide Information Based on Object-Oriented Approach and Cause Analysis in Shuicheng, China. *Remote Sens.* **2022**, *14*, 502. [CrossRef]
9. Chen, T.; Trinder, J.C.; Niu, R. Object-oriented landslide mapping using ZY-3 satellite imagery, random forest and mathematical morphology, for the Three-Gorges Reservoir, China. *Remote Sens.* **2017**, *9*, 333. [CrossRef]
10. Vaduva, C.; Gavat, I.; Datcu, M. Deep learning in very high resolution remote sensing image information mining communication concept. In Proceedings of the 20th European Signal Processing Conference (EUSIPCO), Bucharest, Romania, 27–31 August 2012; pp. 2506–2510.
11. Peña, J.M.; Gutiérrez, P.A.; Hervás-Martínez, C.; Six, J.; Plant, R.E.; López-Granados, F. Object-based image classification of summer crops with machine learning methods. *Remote Sens.* **2014**, *6*, 5019–5041. [CrossRef]
12. Krizhevsky, A.; Sutskever, I.; Hinton, G.E. Imagenet classification with deep convolutional neural networks. *Commun. ACM* **2017**, *60*, 84–90. [CrossRef]
13. Long, J.; Shelhamer, E.; Darrell, T. Fully convolutional networks for semantic segmentation. In Proceedings of the IEEE Conference on Computer Vision and Pattern Recognition, Boston, MA, USA, 7–12 June 2015; pp. 3431–3440.
14. Zhu, X.X.; Tuia, D.; Mou, L.; Xia, G.S.; Zhang, L.; Xu, F.; Fraundorfer, F. Deep learning in remote sensing: A comprehensive review and list of resources. *IEEE Geosci. Remote Sens. Mag.* **2017**, *5*, 8–36. [CrossRef]
15. Ghorbanzadeh, O.; Tiede, D.; Dabiri, Z.; Sudmanns, M.; Lang, S. Dwelling extraction in refugee camps using cnn—First experiences and lessons learnt. In Proceedings of the ISPRS TC I Mid-term Symposium "Innovative Sensing—From Sensors to Methods and Applications" Conference, Karlsruhe, Germany, 10–12 October 2018.
16. He, K.; Zhang, X.; Ren, S.; Sun, J. Deep residual learning for image recognition. In Proceedings of the IEEE Conference on Computer Vision and Pattern Recognition, Las Vegas, NV, USA, 27–30 June 2016; pp. 770–778.
17. Natarajan, A.; Bharat, K.; Kaustubh, G.R.; Moharir, M.; Srinath, N.; Subramanya, K. An Approach to Real Time Parking Management using Computer Vision. In Proceedings of the 2nd International Conference on Control and Computer Vision, Jeju Island, Korea, 15–18 June 2019; pp. 18–22. [CrossRef]
18. Ren, S.; He, K.; Girshick, R.; Sun, J. Faster r-cnn: Towards real-time object detection with region proposal networks. *Adv. Neural Inf. Process. Syst.* **2015**, *28*, 91–99. [CrossRef]
19. Chen, L.C.; Papandreou, G.; Kokkinos, I.; Murphy, K.; Yuille, A.L. Deeplab: Semantic image segmentation with deep convolutional nets, atrous convolution, and fully connected crfs. *IEEE Trans. Pattern Anal. Mach. Intell.* **2017**, *40*, 834–848. [CrossRef]
20. Zhao, H.; Shi, J.; Qi, X.; Wang, X.; Jia, J. Landslide Detection of Hyperspectral Remote Sensing Data Based on Deep Learning with Constrains. *IEEE J. Sel. Top. Appl. Earth Obs. Remote Sens.* **2019**, *12*, 5047–5060. [CrossRef]
21. Ghorbanzadeh, O.; Blaschke, T.; Gholamnia, K.; Meena, S.R.; Tiede, D.; Aryal, J. Evaluation of different machine learning methods and deep-learning convolutional neural networks for landslide detection. *Remote Sens.* **2019**, *11*, 196. [CrossRef]
22. Prakash, N.; Manconi, A.; Loew, S. Mapping landslides on EO data: Performance of deep learning models vs. traditional machine learning models. *Remote Sens.* **2020**, *12*, 346. [CrossRef]
23. Zhu, Q.; Chen, L.; Hu, H.; Xu, B.; Zhang, Y.; Li, H. Deep Fusion of Local and Non-Local Features for Precision Landslide Recognition. *arXiv* **2020**, arXiv:2002.08547.
24. Ji, S.; Yu, D.; Shen, C.; Li, W.; Xu, Q. Landslide detection from an open satellite imagery and digital elevation model dataset using attention boosted convolutional neural networks. *Landslides* **2020**, *17*, 1337–1352. [CrossRef]
25. Liu, P.; Wei, Y.; Wang, Q.; Chen, Y.; Xie, J. Research on post-earthquake landslide extraction algorithm based on improved U-Net model. *Remote Sens.* **2020**, *12*, 894. [CrossRef]
26. Ju, Y.; Xu, Q.; Jin, S.; Li, W.; Su, Y.; Dong, X.; Guo, Q. Loess Landslide Detection Using Object Detection Algorithms in Northwest China. *Remote Sens.* **2022**, *14*, 1182. [CrossRef]
27. Dai, B.; Wang, Y.; Ye, C.; Li, Q.; Yuan, C.; Lu, S.; Li, Y. A Novel Method for Extracting Time Series Information of Deformation Area of A single Landslide Based on Improved U-Net Neural Network. *Front. Earth Sci.* **2021**, *9*, 1139. [CrossRef]
28. Ullo, S.L.; Mohan, A.; Sebastianelli, A.; Ahamed, S.E.; Kumar, B.; Dwivedi, R.; Sinha, G.R. A new mask R-CNN-based method for improved landslide detection. *IEEE J. Sel. Top. Appl. Earth Obs. Remote Sens.* **2021**, *14*, 3799–3810. [CrossRef]
29. Liu, P.; Wei, Y.; Wang, Q.; Xie, J.; Chen, Y.; Li, Z.; Zhou, H. A research on landslides automatic extraction model based on the improved mask R-CNN. *ISPRS Int. J. Geo-Inf.* **2021**, *10*, 168. [CrossRef]
30. Ghorbanzadeh, O.; Gholamnia, K.; Ghamisi, P. The application of ResU-net and OBIA for landslide detection from multi-temporal sentinel-2 images. In *Big Earth Data*; Taylor & Francis: Abingdon, UK, 2022; pp. 1–26. [CrossRef]
31. Dahmane, M.; Foucher, S.; Beaulieu, M.; Riendeau, F.; Bouroubi, Y.; Benoit, M. Object detection in pleiades images using deep features. In Proceedings of the 2016 IEEE International Geoscience and Remote Sensing Symposium (IGARSS), Beijing, China, 10–15 July 2016; pp. 1552–1555. [CrossRef]
32. Längkvist, M.; Alirezaie, M.; Kiselev, A.; Loutfi, A. Interactive learning with convolutional neural networks for image labeling. In Proceedings of the International Joint Conference on Artificial Intelligence (IJCAI), New York, NY, USA, 9–15 July 2016.

33. Ronneberger, O.; Fischer, P.; Brox, T. U-net: Convolutional networks for biomedical image segmentation. In Proceedings of the International Conference on Medical Image Computing and Computer-Assisted Intervention, Munich, Germany, 5–9 October 2015; pp. 234–241. [CrossRef]
34. Chen, L.C.; Zhu, Y.; Papandreou, G.; Schroff, F.; Adam, H. Encoder-decoder with atrous separable convolution for semantic image segmentation. In Proceedings of the European Conference on Computer Vision (ECCV), Munich, Germany, 8–14 September 2018; pp. 801–818. [CrossRef]
35. Zhao, H.; Shi, J.; Qi, X.; Wang, X.; Jia, J. Pyramid scene parsing network. In Proceedings of the IEEE Conference on Computer Vision and Pattern Recognition, Honolulu, HI, USA, 21–16 July 2017; pp. 2881–2890.

applied sciences

MDPI

Article

Chinese Named Entity Recognition of Geological News Based on BERT Model

Chao Huang, Yuzhu Wang *, Yuqing Yu, Yujia Hao, Yuebin Liu and Xiujian Zhao

School of Information Engineering, China University of Geosciences, Beijing 100083, China; huangchao@cugb.edu.cn (C.H.); yuyq@cugb.edu.cn (Y.Y.); haoyj@cugb.edu.cn (Y.H.); liuyb@cugb.edu.cn (Y.L.); zhaoxj@cugb.edu.cn (X.Z.)

* Correspondence: wangyz@cugb.edu.cn

Abstract: With the ongoing progress of geological survey work and the continuous accumulation of geological data, extracting accurate information from massive geological data has become increasingly difficult. To fully mine and utilize geological data, this study proposes a geological news named entity recognition (GNNER) method based on the bidirectional encoder representations from transformers (BERT) pre-trained language model. This solves the problems of traditional word vectors that are difficult to represent context semantics and the single extraction effect and can also help construct the knowledge graphs of geological news. First, the method uses the BERT pre-training model to embed words in the geological news text, and the dynamically obtained word vector is used as the model's input. Second, the word vector is sent to a bidirectional long short-term memory model for further training to obtain contextual features. Finally, the corresponding six entity types are extracted using conditional random field sequence decoding. Through experiments on the constructed Chinese geological news dataset, the average F1 score identified by the model is 0.839. The experimental results show that the model can better identify news entities in geological news.

Keywords: BERT; named entity recognition; geological news; CRF

Citation: Huang, C.; Wang, Y.; Yu, Y.; Hao, Y.; Liu, Y.; Zhao, X. Chinese Named Entity Recognition of Geological News Based on BERT Model. *Appl. Sci.* **2022**, *12*, 7708. https://doi.org/10.3390/app12157708

Academic Editor: Douglas O'Shaughnessy

Received: 8 July 2022
Accepted: 28 July 2022
Published: 31 July 2022

1. Introduction

With the rapid development of society, artificial intelligence has been applied in all aspects of our lives [1]. In recent years, artificial intelligence has developed rapidly in various fields. Taking natural language processing (NLP) as an example, it is applied in many fields, such as information extraction, question answering systems, machine translation, and text classification [2]. Named entity recognition (NER) is an important field in NLP technology, and many studies on NLP are based on it. Currently, research on NER is mainly concentrated in the fields of finance and medical care and has not yet been seen in the field of geological news.

In early research on NER, rule-based and dictionary-based methods were mainly used [3–5]. Later, with the development of machine learning, machine learning methods were used to solve NER tasks. Common methods include hidden Markov models [6,7], maximum entropy models [8,9] and conditional random field (CRF) models [10], among others. In recent years, with the development of computer technology and deep learning, the method of using deep learning models has become a trend for solving NLP problems. In the field of geology, some scholars have used deep learning-based models for NER tasks and have achieved good results. Zhang et al. [11] proposed a geological entity recognition model based on a deep belief network, which achieved good entity recognition results on a small-scale corpus, and each evaluation index (P, R, F1) reached 90% and above. Liu et al. [12] proposed an improved lattice long- and short-term memory (LSTM) model based on a bidirectional long short-term memory conditional random field (BiLSTM-CRF). The proposed model is based on LSTM [13] and achieved good results for named entities in the coal mining field, with an F1 score of 94.04% and an improvement of 2.1% based on the original BiLSTM-CRF.

There is a huge amount of geological news texts that contain a large amount of information. Accurately identifying effective information from them can provide important data support for related geological survey work. However, traditional manual extraction methods have problems such as high time consumption and low accuracy. As the scale of geological news text data increases, the extraction becomes more and more difficult. Therefore, it is important to realize the automatic extraction of geological news information entities, which is also the basic work of geological news knowledge graph construction.

Geological news text data are complex and contain many types of data. Related entities include time, geographic location, organization, job title, event, etc. Geological news texts are different from common news texts. Because they are news related to geology, the names of related entities have obvious characteristics, such as a professional background and application behavior—for example, an organization entity (China Natural Resources Airborne Geophysical and Remote Sensing Center). In addition, the text also has polysemy and entity nesting problems, such as China referring to either a geographical location or a country. The China Geological Survey of the Ministry of Natural Resources contains an organizational entity (Ministry of Natural Resources), a geographical location entity (China), and an organizational entity (Geological Survey). At present, there is no public dataset in the field of geological journalism. Therefore, it is challenging to construct a corpus of geological journalism before carrying out the NER task.

To accurately extract entities from geological news, this study proposes a model that combines a bidirectional encoder representations from transformers (BERT) pre-trained model and a BiLSTM-CRF model for geological news named entity recognition (GNNER). The model first uses the pre-trained word vector model BERT for semantic extraction. Compared to the traditional word vector, the GNNER Word2vec, BERT [14] can better represent semantic information in different contexts to solve the polysemy problem. After obtaining the output of the BERT model, part-of-speech analysis and chunking analysis features are added to help the model identify entity boundaries. Finally, the word vector is sent to the BiLSTM model for further training. The results of the BiLSTM model are modified using CRF, outputting the labeled sequence with the highest score. Based on the geological news texts of the China Geological Survey and according to the characteristics of geological news texts, time, name, geographic location, organization, and other information are extracted. The experimental results show that the model could better identify the entities in geological news.

The main contributions of this paper are summarized as follows:

(1) GNNER was based on the BERT model, integrating a variety of different models and extracting various types of entities from the constructed geological news corpus.
(2) This research used crawler technology to obtain geological news texts from the China Geological Survey Bureau, preprocessed the data, including long text segmentation, data cleaning, and removal of uncommon punctuation marks, and used the "BIO" named entity labeling method to label the texts to create a dataset of a certain scale in the field of geological news.
(3) The BERT-BiLSTM-CRF model was used to conduct a comparative experiment with the other five models on the geological news dataset, analyze the quality of the six models, and discuss the effects of geological news entity type, number of labels, and model hyperparameters on model evaluation.

The rest of the paper is structured as follows: Section 2 introduces the related model design methods and the dataset construction process; Section 3 presents the experimental results; and Section 4 discusses the experimental results and directions for future research.

2. Materials and Methods

2.1. Word2vec

Before using the deep learning model to solve the NLP problem, we needed to convert the language data type into a data type that the neural network could handle. Word embedding technology was developed because of this requirement.

The Word2vec model was developed by Tomas Mikolov [15] et al. in 2013. It is an efficient model for training word vectors. Unlike the traditional language model, Word2vec assumes that there is a relationship between similar words in a sentence. It has two models: the skip-gram and CBOW. The model structure of Word2vec is shown in Figure 1. The skip-gram uses the current word to predict nearby words, whereas CBOW uses nearby words to predict the current word.

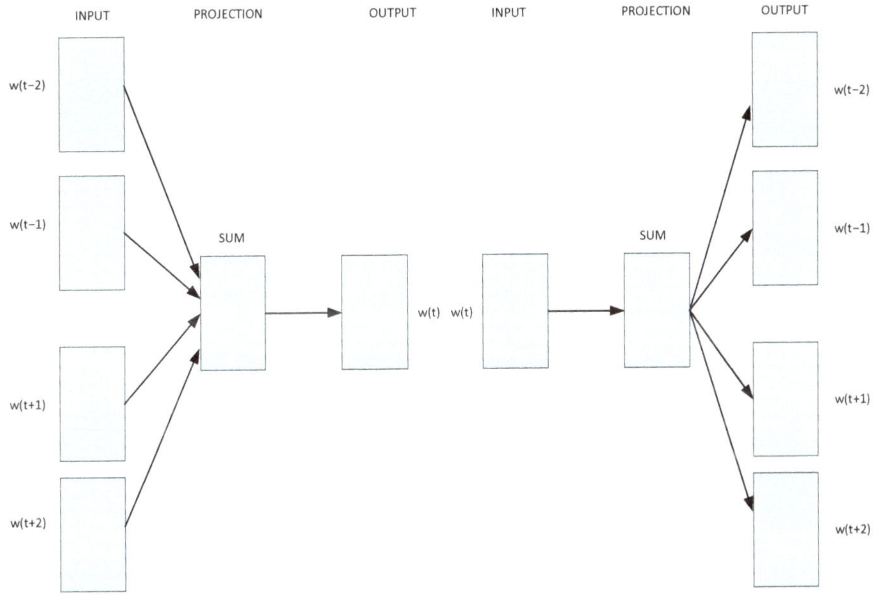

Figure 1. CBOW model (**left**) and skip-gram model (**right**).

Although the Word2vec model has achieved good results in word embedding, it also has some problems. For example, the word vectors trained using this method are fixed and cannot change the meaning of words in different contexts.

2.2. BiLSTM-CRF

Proposed by Lample [16] et al. and based on the LSTM-CRF model, the BiLSTM-CRF model is a deep learning model that integrates feature engineering and serialization. The model structure is shown in Figure 2. It is mainly divided into a three-layer structure of a word vector input layer, a BiLSTM layer, and a CRF layer. The experimental process can be divided into three steps. First, the input of the model is a sequence of word vectors. Second, the probability vector of the corresponding label of each word is output on the BiLSTM layer. Finally, the result is corrected through the CRF, and the label sequence with the highest probability is output. These three parts are explained in detail below.

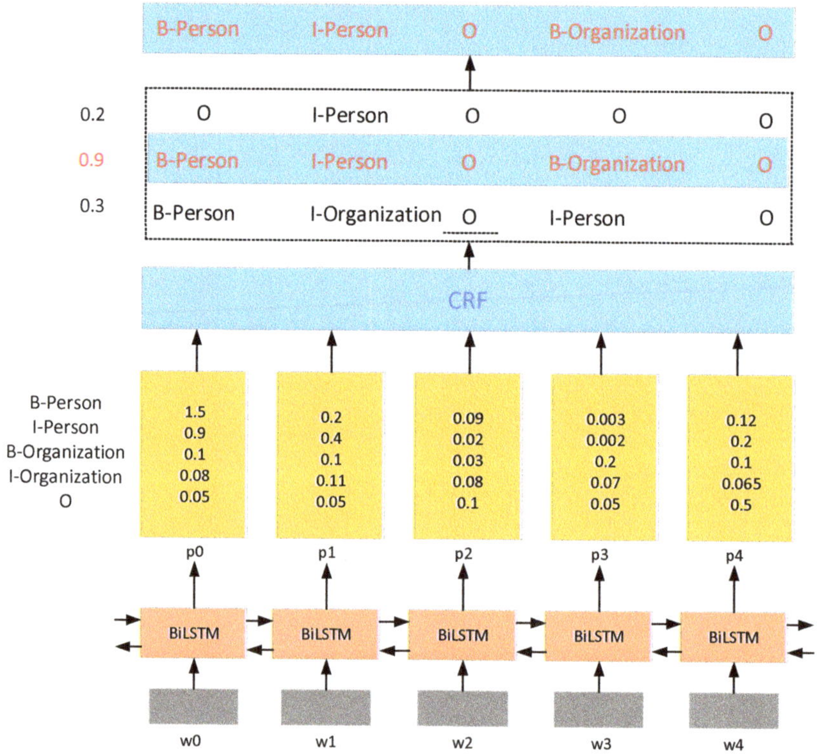

Figure 2. The structure of the BiLSTM-CRF model.

The first part is the word vector input layer, where the sentence is input. Assuming that the sentence is X, X is composed of n words $(X_1, X_2, \ldots, X_n)$, denoted as $X = (X_1, X_2, \ldots, X_n)$. Each word in the sentence is mapped into a word vector using the word embedding method, corresponding to w_0, w_1, w_2, w_3, w_4, etc. In the figure, the matrix formed by these word vectors is used as the input of the next layer.

The second part is the BiLSTM layer, which has two hidden layers: a forward LSTM layer and a backward LSTM layer. The input word vector goes through the two hidden layers, and the results of these hidden layers are spliced together to determine the state of the final hidden layer. Based on the state of the final hidden layer, a linear layer is connected to perform a mapping operation to map the hidden layer matrix from n dimensions to K dimensions, where n is the dimension of the matrix and K is the number of labels. Finally, the BiLSTM layer will output the resulting matrix, which is denoted as $P = (p_0, p_1, \ldots, p_n)$, where p_i is the vector composed of the emission scores of w_i corresponding to each label, e.g., $p_0 = [1.5, 0.9, 0.1, 0.08, 0.05]$. The matrix P serves as the input matrix of the CRF layer.

The last part is the CRF layer, which adds constraints between the labels and reduces the number of invalid predicted labels. Although the Softmax function outputs the label with the maximum probability corresponding to the word, the output labels are independent of each other. This means that the sequence is prone to unreasonable situations, resulting in a decrease in the accuracy rate, such that the time and organization entity of the "I" label may become adjacent. However, we know that according to the "BIO" labeling method, the entity tag must begin with "B". Thus, the "I" label of the time and organization entity cannot be adjacent.

After the word vector passes through the BiLSTM and CRF layers, the score of the final sequence consists of two parts: the emission score of the BiLSTM layer and the transfer score of the CRF layer, as shown in Equation (1). The X on the left side of the equal sign represents the scoring sequence, and the first part on the right side of the equal sign P_{i,y_i} represents the emission score of the y_i label in the ith word vector. The second part A_{y_{i-1},y_i} represents the emission score from y_{i-1} label to the y_i label.

$$socre(X,y) = \sum_{i=1}^{n} P_{i,y_i} + \sum_{i=1}^{n+1} A_{y_{i-1},y_i} \tag{1}$$

After calculating the score for each possible sequence, it is normalized using Softmax. The result is shown in Equation (2), which $Y_{(x)}$ represents all possible labeled sequences.

$$p(y|X) = \frac{e^{socre(X,y)}}{\sum \tilde{y} \in Y_{(x)} e^{socre(X,\tilde{y})}} \tag{2}$$

In the model training process, the log-likelihood function is used to optimize the model. The result is shown in Equation (3).

$$\log(p(y|X)) = \log(\frac{e^{socre(X,y)}}{\sum \tilde{y} \in Y_{(x)} e^{socre(X,\tilde{y})}}) = score(X,y) - \log \sum_{\tilde{y} \in Y(x)} e^{socre(X,\tilde{y})} \tag{3}$$

Finally, the Viterbi algorithm [17] is used to decode the hidden state sequence to obtain the optimal label sequence. The result is shown in Equation (4).

$$y^* = \mathrm{arg}max_{\tilde{y} \in Y_{(x)} socre(X,\tilde{y})} \tag{4}$$

2.3. BERT

In Section 2.1, we discussed the need to convert the data type into a data type that the neural network can handle when dealing with NLP problems; thus, word embedding technology is needed. However, traditional word embedding technology has some problems, such as its inability to solve polysemy and dynamically optimize specific tasks. To solve these problems, Jacob Devlin et al. proposed a new pre-training model called the BERT model in 2018. BERT is a deep bidirectional language representation model pre-trained on a corpus consisting of a large number of books and Wikipedia, and its main structure is the encoder part of the Transformer model [18].

The input of the BERT model is a token sequence, which is inserted [CLS] at the beginning of each sequence to classify sentences and [SEP] at the end of the sequence to separate different sentences. Each token sequence consists of three parts: token embeddings, segment embeddings, and position embeddings.

The Transformer model was developed by Google's Vaswani et al. in 2017. This model efficiently realizes the parallelization of non-serialized models, which can greatly improve computational efficiency. Figure 3 shows a structural diagram of the Transformer model. As the structure of the BERT model is mainly the encoder part, the composition and principles of the encoder part are explained as follows.

(1) Input of Transformer

The Transformer model trains all words in the sequence at the same time. To identify the position information of each word in the sequence, it is necessary to add a position encoding (PositionEncoding) to each word vector (EmbeddingLookup (X)), as shown in Equation (5).

$$\mathrm{X} = \mathrm{EmbeddingLookup}(X) + \mathrm{PositionalEncoding} \tag{5}$$

(2) Self-attention mechanism

Unlike the attention mechanism [19], the self-attention mechanism calculates the relationship between the elements in the input or output sequence, which is an improved method based on the attention mechanism. In the calculation process of the self-attention mechanism, the matrix Query, Key, and Value need to be used. From the perspective of the information retrieval system, Query is the input information, Key is the content information matching Query, and Value is the information itself. The calculation process is then described in detail. Query, Key, and Value are denoted as Q, K, and V, respectively.

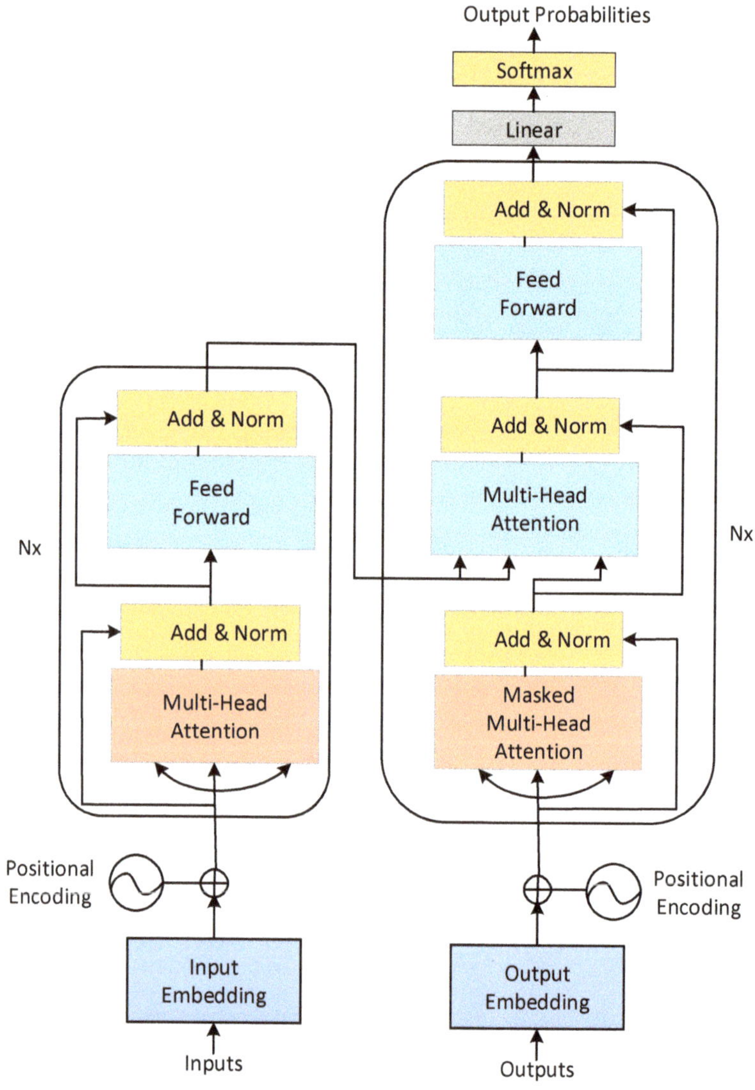

Figure 3. The structure of Transformer [18].

The input of the self-attention mechanism is the matrix X. Q, K, and V are obtained by the linear transformation of X, as shown in Equations (6)–(8), where W_Q, W_k, W_v are the three auxiliary matrices. The word vector matrix X is multiplied by these auxiliary matrices to obtain the corresponding Q, K, and V values for each item in the sequence. The Q of the current item is multiplied by the K of each item in the sequence to determine the relationship between the two. After scaling and normalizing the product using Softmax, it is multiplied by V, and each V is added to obtain the feature representation of the current item. In Equation (9), d_k is the dimension of the Q and K vectors.

$$Q = Linear(X) = XW_Q \tag{6}$$

$$K = Linear(X) = XW_k \tag{7}$$

$$V = Linear(X) = XW_v \tag{8}$$

$$X_{attention} = SelfAtteention(Q, K, V) = softmax\left(\frac{QK^T}{\sqrt{d_k}}\right)V \tag{9}$$

(3) Multi-head mechanism

In the self-attention mechanism, each item in the input sequence corresponds to a set of feature expressions: Query, Key, and Value. Conversely, the so-called multi-head mechanism establishes multiple sets of auxiliary matrices in the Transformer model and multiplies them by the word vector input matrix X to obtain multiple sets of the Query, Key, and Value values. Therefore, each item in the sequence has multiple sets of feature expressions. Multiple sets of feature expressions are spliced together, and dimensionality reduction is performed using a fully connected layer.

(4) Summation and normalization

A residual connection is also required to ensure a better feature extraction effect. The residual connection adds the vector after the self-attention mechanism and the multi-head mechanism to the original input vector, as shown in Equations (10) and (11). It is necessary to normalize the hidden layer to speed up convergence.

$$X_{attention} = X + X_{attention} \tag{10}$$

$$X_{attention} = LayerNorm(X_{attention}) \tag{11}$$

After extensive training, the BERT model can be applied to various natural language processing tasks. This paper uses the BERT model instead of Word2vec to obtain word vectors that can better integrate context information and improve the accuracy of named entity recognition.

In addition, this paper also uses two improved models based on BERT for experiments, namely distilled BERT (DistilBERT) and robustly optimized BERT approach (RoBERTa). DistilBERT is a distilled version of BERT proposed by Victor Sanh [20] et al., which is smaller, faster and cheaper than the BERT model. RoBERTa is a robustly optimized BERT pre-training approach proposed by Liu [21] et al. By improving BERT, these two models enable BERT to achieve high performance on large datasets.

2.4. Model Design

In the NLP task, as the BERT model has achieved good results, more and more people have begun to combine BERT with deep learning models for NER tasks. In this study, we introduce the BERT model based on the BiLSTM-CRF model and design the BERT-BiLSTM-CRF model, which is used to identify geological news entities. As shown in Figure 4, the structure of the model is mainly divided into three layers from bottom to top: the BERT layer, BiLSTM layer, and CRF layer. First, the input of the BERT model is the superposition of each word vector, including the sentence vector and the position vector. The word vector can obtain the text context features after the encoding layer of the Transformer. Therefore, the word vector output after BERT training can also be remarkable and effectively integrate the article features. Second, the BERT output result is used as the input of the BiLSTM layer, and the context information can be better integrated using the two-layer LSTM neural network before and after. Finally, the labeling sequence output by the BiLSTM layer goes through the CRF layer, and the labeling sequence is corrected by the state transition matrix. The optimal labeling sequence is finally output.

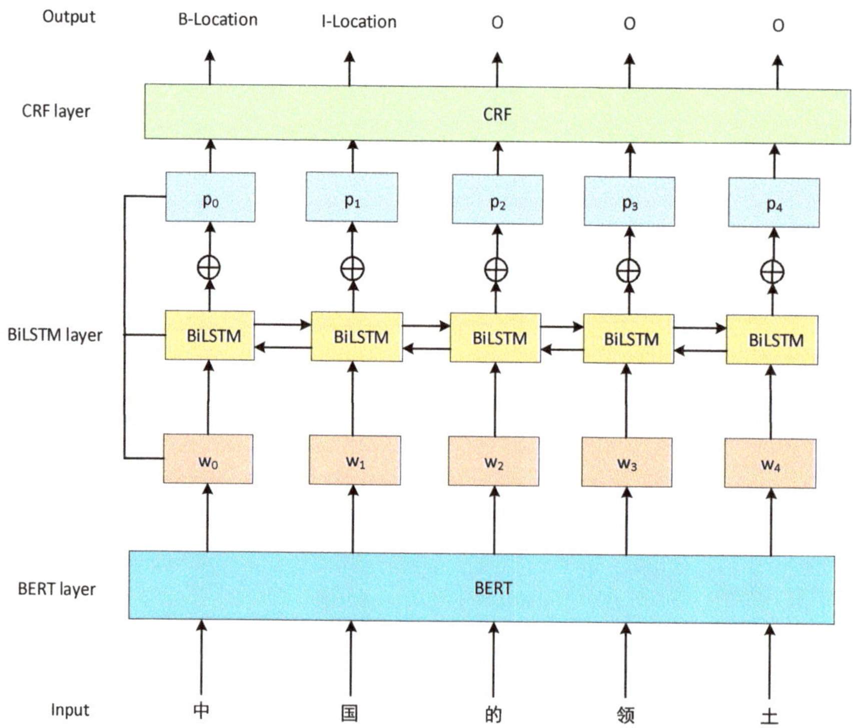

Figure 4. The structure of the BERT-BiLSTM-CRF model. The Chinese characters in the figure show the first to fifth characters of the territory of China expressed in Chinese.

2.5. Data Processing and Experimental Setup

2.5.1. Data Source and Pre-Processing

The original data of this paper are a total of 2200 geological news texts obtained from the China Geological Survey through web crawlers with about one million words. As the data have much noise, it was necessary to preprocess and clean the data to reduce the effect on the experimental accuracy. First, special symbols, such as some non-Chinese, non-English, and non-digital symbols, were removed from the text. Second, the text contains elements that have nothing to do with geological news. This kind of content has obvious signs before and after the text, so that it can be used mainly for filtering using a regular expression. News texts that were too long needed to be segmented. In this study, news texts were segmented according to the priority of punctuation. The preprocessed dataset consists of Chinese and English characters, numbers, punctuation marks and spaces, and the length of each sentence after segmentation does not exceed the max_len set in the experimental parameters.

2.5.2. Text Corpus Annotation

Sequence labeling is the most important step in the construction of datasets. There are many popular labeling methods, and the "BIO" labeling method is used in this paper. In the "BIO" notation method, a paragraph is usually marked as "B-X," "I-X," and "O;". "B" means "begin," which in Chinese refers to the Chinese character at the beginning of a named entity; "I" means "Inside," which refers to the middle and end parts of the named entity; and "O" means "Other," which refers to the non-named entity part.

The task of labeling named entities is cumbersome. To reduce the intensity of the task, this study uses YEDDA [22] as an auxiliary tool for labeling corpora. After the text is marked, according to statistics, 21,323 entities were marked, including 6 entity types: name, time, geographic location, job title, organization, and event. Through the statistics of the number of various entities in the labeled corpus, the specific labeling situation is shown in Table 1. In the labeling process, we labelled complex entities with multiple entities one by one. For example, the entities "Ministry of Natural Resources China Geological Survey", "Ministry of Natural Resources", "China" and "Geological Survey" were annotated as an organization, geographic location, and organization, respectively.

Table 1. The number of various entities in the annotated corpus.

Tags	Type	Number
TIM	Time	3492
ORG	Organization	5630
POS	Job title	1050
EVE	Event	2032
LOC	Geographic location	7702
PER	Name	1417

2.5.3. Experimental Environment and Parameters

The BERT model requires strong computing power during the training process, so it has certain requirements for computer hardware. Table 2 presents the relevant information on the hardware and software used in this experiment.

Table 2. Experimental environment. CUDA is a parallel computing platform and programming model invented by NVIDIA, headquartered in Santa Clara, California, USA. Python is a programming language designed by Guido van Rossum in the Netherlands. Tensorflow is a symbolic math system developed by Google's artificial intelligence team, Google Brain. Pytorch is an open source Python machine learning library launched by the Facebook Artificial Intelligence Research Institute. Numpy is an extension tool for numerical computing developed by Jim Hugunin and other collaborators.

Category	Configuration
Hardware	GPU: 4*NVIDIA Tesla K80 OS: CentOS 8.3 Video memory: 11 GB GDDR6
Software	CUDA: 11.4 Python: 3.6 Tensorflow: 1.14.0 Pytorch: 1.4.0 Numpy: 1.19.2

In the training process of the model, the setting of hyperparameters influences the training effect. To exclude the effects of different hyperparameters on the experiment, fixed hyperparameters were used to train different models. Table 3 shows several important parameters used in the model training process. Among these parameters, an epoch is a process of training the training set once, max_len is the length of the maximum sequence, batch_size is the amount of data obtained in one training process, learning_rate is the learning rate, and drop_rate is set to prevent overfitting of the neural network.

Table 3. Model parameters.

Hyper-Parameter	Parameter Values
Epochs	8
max_len	128
batch_size	16
learning_rate	3×10^{-5}
drop_rate	0.5

3. Results

3.1. Experimental Evaluation Indicators

In the process of NER, evaluation indicators are needed to evaluate the quality of the model. In this study, the three evaluation indicators used in all experiments are precision rate (P), recall rate (R), and F1 score (F1).

3.2. Comparison of Different Models

We divided the labeled dataset into the training set, validation set, and test set according to a ratio of 8:1:1, with 17,124, 2056, and 2143 entities, respectively. In the NER task of the geological news dataset, we trained the dataset on six different models and tested it on the test set. The experimental results show that the model has achieved good results in GNNER. Table 4 presents relevant information on the precision rate, recall rate, and F1 scores of the six models for six categories of entities: name, time, geographic location, job title, organization, and event.

Table 4. P, R, and F1 scores of six types of entities on six models. The numbers in bold font are the three indicators (P, R, F1) with the highest corresponding scores in the experiment.

Model	Eval	TIM	ORG	POS	EVE	LOC	PER	Avg
BERT	P	0.863	0.820	0.691	0.820	0.802	0.887	0.819
	R	0.827	0.844	0.827	0.796	0.835	0.806	0.829
	F	0.845	0.832	0.753	0.808	0.818	0.844	0.824
DistilBERT	P	0.847	0.796	0.721	0.793	0.787	0.864	0.808
	R	0.804	0.852	0.803	0.812	0.824	0.816	0.821
	F	0.825	0.823	0.760	0.802	0.805	0.839	0.814
RoBERTa	P	0.865	0.828	0.725	0.815	0.808	0.876	0.823
	R	0.823	0.841	0.814	0.821	0.827	0.829	0.834
	F	0.843	0.834	0.767	0.818	0.817	0.852	0.828
BiLSTM-CRF	P	0.893	0.876	0.864	0.563	0.812	0.854	0.814
	R	0.765	0.803	0.760	0.697	0.687	0.616	0.728
	F	0.824	0.838	0.809	0.623	0.744	0.716	0.768
BERT-CRF	P	0.841	0.837	0.733	0.803	0.849	0.878	0.838
	R	0.860	0.847	0.811	0.808	0.841	0.840	**0.841**
	F	0.850	0.842	0.770	0.805	0.845	0.859	**0.839**
BERT-BiLSTM-CRF	P	0.844	0.844	0.739	0.853	0.843	0.827	**0.839**
	R	0.835	0.846	0.863	0.838	0.838	0.811	0.838
	F	0.839	0.845	0.796	0.846	0.840	0.819	0.838

By analyzing the above experimental results, we can draw the following conclusions:

(1) The six models adopted in the experiment have achieved good results in the geological news text NER task.

(2) In the geological news text NER task, the F1 scores of the BERT, DistilBERT, RoBERTa, BERT-CRF, and BERT-BiLSTM-CRF models are 0.824, 0.814, 0.828, 0.839, and 0.838, respectively. Compared to BiLSTM-CRF, the F1 scores increase by 5.6%, 4.6%, 6%, 7.1%, and 7%. This shows that as the BERT pre-training model can understand the contextual information of the text well and solve the polysemy problem, it has a good effect on the named entity recognition task.

(3) The improved DistilBERT and RoBERTa models based on BERT achieve F1 scores of 0.814 and 0.828, respectively. Compared to the BERT model, the entity recognition effect of the DistilBERT model is slightly worse, while the RoBERTa model is better.

(4) In the geological news text NER task, the P, R, and F1 scores of the BERT-CRF model improve by 1.9%, 1.2%, and 1.5%, respectively, compared to the BERT model because of a mutual constraint relationship between the tags (e.g., the tag of an entity can only start with "B" but not "I"). It can be seen after adding the CRF layer that CRF can

deal with the mutual constraint relationship between the tags and effectively solve the problem of inconsistent sequence tags.
(5) The P, R, and F1 scores of BERT-CRF are 0.838, 0.841, and 0.839, respectively, which are the best among all models. Compared to the BERT-BiLSTM-CRF model, which introduced the BiLSTM layer, the two achieved comparable results in the NER task of geological news texts. The reason for this is that the BERT model itself is effective in feature extraction, and the BiLSTM layer is introduced based on the BERT-CRF model. Overfitting occurs after this layer is trained, resulting in a decreased effect.

3.3. Effect of Entity Type and Quantity

Figure 5 shows the F1 scores of the six models in the six entity categories in the form of a bar chart. Using the same model, there is a gap in the recognition effect for the different entity categories. The recognition effect is better for the entities of time, organization, name, and geographic location because the number of these entities in the corpus is large, their contextual information is more abundant, and the text features are more obvious. The recognition effect of job titles and events is poor. Two reasons account for this result: (1) the number of these two types of entities is small, especially job titles, which leads to insufficient contextual information for the neural network to learn; (2) the entities of the event class are usually nested entities, and geographic location and organization entities often appear, increasing the difficulty of identification.

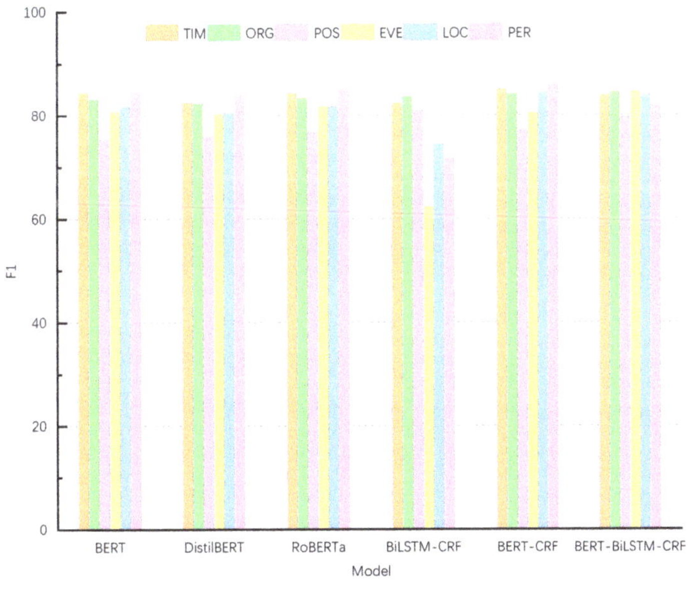

Figure 5. F1 scores of six types of entities on six models.

3.4. Influence of Model Hyperparameters

The hyperparameters need to be set before the model is trained. The setting of hyperparameters plays a role in the training effect of the model. For different models and datasets, experiments are often required to find the most suitable hyperparameters. In the experiment, we mainly discussed the effects of the learning rate and training times on the model training effect. The F1 score of the model is used as the condition for parameter evaluation. When the F1 score is the largest, this means that the parameter is the best parameter. In deep learning, the learning rate is a common hyperparameter. This setting not only determines whether the objective function can converge but also affects the speed of convergence. When the setting is too small, the convergence speed may be too slow; when the setting is too large, it may lead to non-convergence. Therefore, choosing an appropriate learning rate is critical during model training. Based on all models, the experiment was performed by changing the learning rate of the model. Table 5 shows the experimental results for different learning rates. The F1 scores of all models are the largest when the learning rate is 3×10^{-5}, which means that the entity recognition effect of the model is the best under this parameter condition.

Table 5. P, R and F1 scores of the model under different learning rates. The numbers in bold font are the three indicators (P, R, F1) with the highest corresponding scores in each set of experiments.

Model	Learning_Rate	P	R	F
BiLSTM-CRF	1×10^{-5}	0.795	0.725	0.758
	2×10^{-5}	0.804	**0.732**	0.766
	3×10^{-5}	0.814	0.728	**0.768**
	4×10^{-5}	**0.817**	0.721	0.766
	5×10^{-5}	0.806	0.725	0.763
BERT	1×10^{-5}	0.805	0.823	0.814
	2×10^{-5}	0.814	0.827	0.821
	3×10^{-5}	0.819	**0.829**	**0.824**
	4×10^{-5}	**0.823**	0.819	0.821
	5×10^{-5}	0.817	0.822	0.819
DistilBERT	1×10^{-5}	0.798	0.812	0.805
	2×10^{-5}	0.803	0.814	0.808
	3×10^{-5}	0.808	**0.821**	**0.814**
	4×10^{-5}	**0.812**	0.811	0.811
	5×10^{-5}	0.805	0.818	0.811
RoBERTa	1×10^{-5}	0.809	0.816	0.812
	2×10^{-5}	0.817	0.822	0.819
	3×10^{-5}	0.823	**0.834**	**0.828**
	4×10^{-5}	**0.826**	0.828	0.827
	5×10^{-5}	0.821	0.827	0.824
BERT-BiLSTM-CRF	1×10^{-5}	0.821	0.823	0.822
	2×10^{-5}	0.832	0.833	0.832
	3×10^{-5}	**0.839**	0.838	**0.838**
	4×10^{-5}	0.834	**0.838**	0.836
	5×10^{-5}	0.834	0.825	0.829
BERT-CRF	1×10^{-5}	0.823	0.819	0.821
	2×10^{-5}	0.829	0.832	0.831
	3×10^{-5}	**0.838**	**0.841**	**0.839**
	4×10^{-5}	0.836	0.838	0.837
	5×10^{-5}	0.832	0.824	0.828

Figure 6 is a line graph showing the effect of epoch times on the F1 score. As shown in the figure, the abscissa is the epoch value, and the ordinate is the F1 score. The F1 scores of the six models all increase with the increase in the number of training rounds. First, the F1 score of the BiLSTM-CRF model in the first few rounds is much lower than that of the BERT, DistilBERT, RoBERTa, BERT-CRF, and BERT-BiLSTM-CRF models. Second, as the number of training increases, the F1 scores of the six models gradually increase. The F1 score of the BiLSTM-CRF model gradually becomes close to the F1 score of the other five models that introduced BERT. At the 8th epoch, the F1 scores of the BERT-CRF model and the BERT-BiLSTM-CRF model reach the maximum. Finally, the F1 scores of the six models tend to be stable, but the F1 score of the BilSTM-CRF model lags behind those of the other five models.

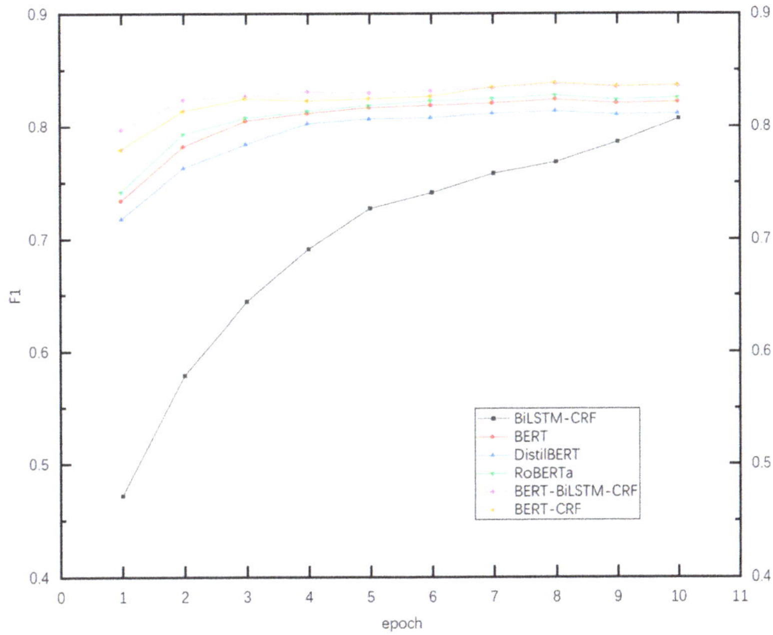

Figure 6. Changes in F1 score with increasing epochs.

4. Conclusions

This study uses a deep learning-based model to perform named entity recognition on geological news texts. As there is no public labeling training dataset for model comparison in the field of geological news, we collected geological news texts from the China Geological Survey. We annotated the data using automatic annotation and manual calibration with the open source annotation tool YEDDA to build a corpus of geological news texts. In past NER research in the geological field, Chen et al. [23] proposed the BERT- BiLSTM-CRF model to perform entity recognition in Chinese mineral texts. In the experiment, eight entity types were extracted, and their F1 scores all exceeded 95%. Xie et al. [24] obtained F1 scores of 94.65% and 95.67% on the MSRA and People's Daily corpora, respectively, in the BERT-BiLSTM-CRF model for named entities. Our experiment compares six models on the constructed geological news dataset. The F1 score of the BERT-BiLSTM-CRF model is 0.838, which achieves a high entity recognition effect, and the F1 score of the BiLSTM-CRF model is 0.768, which is less efficient than the BERT-based models. Compared to the previous study, we annotated 21,323 entities, trained based on the geological news dataset, extracted six entity types, and the F1 score of the model reached 0.839. The experimental results show that the proposed model can also achieve a good entity recognition effect in the field of geological news.

Although this research has achieved good results in NER tasks in the field of geological news, there are still some shortcomings and areas that can be further improved. First, the geological news corpus constructed in this study is relatively small, resulting in too few job titles and event entities, thus influencing the effect of entity recognition. Therefore, the dataset should be expanded in future studies. Second, as the process is cumbersome, it is inevitable that some labeling errors will occur. Moreover, the original model should be improved to enhance its performance and improve the entities' recognition effect. The information extracted from the geological news text can be applied to constructing geological news knowledge graphs.

Author Contributions: Conceptualization, Y.W.; methodology, C.H. and Y.W.; software, C.H.; validation, Y.H., Y.L. and X.Z.; data curation, C.H. and Y.Y.; writing—original draft preparation, C.H.; writing—review and editing, Y.W., Y.Y. and Y.H.; supervision, Y.W. All authors have read and agreed to the published version of the manuscript.

Funding: This work was supported by the China University of Geosciences (Beijing) College Students' Innovation and Entrepreneurship Training Program under Grant X202211415100.

Institutional Review Board Statement: Not applicable.

Informed Consent Statement: Not applicable.

Data Availability Statement: Data sharing is not applicable.

Conflicts of Interest: The authors declare no conflict of interest. The funders had no role in the design of the study, in the collection, analyses, or interpretation of data, in the writing of the manuscript, or in the decision to publish the results.

References

1. Goralski, M.A.; Tan, T.K. Artificial intelligence and sustainable development. *Int. J. Manag. Educ.* **2020**, *18*, 100330. [CrossRef]
2. Khurana, D.; Koli, A.; Khatter, K.; Singh, S. Natural language processing: State of the art, current trends and challenges. *arXiv* **2017**, arXiv:1708.05148. [CrossRef]
3. Shaalan, K.; Raza, H. Arabic named entity recognition from diverse text types. In Proceedings of the International Conference on Natural Language Processing, Gothenburg, Sweden, 25–27 August 2008; pp. 440–451.
4. Alfred, R.; Leong, L.C.; On, C.K.; Anthony, P. Malay named entity recognition based on rule-based approach. *Int. J. Mach. Learn. Comput.* **2014**, *3*, 300–306. [CrossRef]
5. Shaalan, K.; Raza, H. NERA: Named entity recognition for Arabic. *J. Am. Soc. Inf. Sci. Technol.* **2009**, *60*, 1652–1663. [CrossRef]
6. Todorovic, B.T.; Rancic, S.R.; Markovic, I.M.; Mulalic, E.H.; Ilic, V.M. Named entity recognition and classification using context Hidden Markov Model. In Proceedings of the 2008 9th Symposium on Neural Network Applications in Electrical Engineering, Belgrade, Serbia, 25–27 September 2008; pp. 43–46.
7. Eddy, S.R. What is a hidden Markov model? *Nat. Biotechnol.* **2004**, *22*, 1315–1316. [CrossRef] [PubMed]
8. Och, F.J.; Ney, H. Discriminative training and maximum entropy models for statistical machine translation. In Proceedings of the 40th Annual Meeting of the Association for Computational Linguistics, Philadelphia, PA, USA, 7–12 July 2002; pp. 295–302.
9. Ratnaparkhi, A. Maximum Entropy Models for Natural Language Processing. In *Encyclopedia of Machine Learning and Data Mining*; Springer: New York, NY, USA, 2017; pp. 800–805.
10. Lafferty, J.; McCallum, A.; Pereira, F.C. Conditional random fields: Probabilistic models for segmenting and labeling sequence data. In Proceedings of the 18th International Conference on Machine Learning, San Francisco, CA, USA, 28 June–1 July 2001.
11. Zhang, X.Y.; Ye, P.; Wang, S.; Du, M. Geological entity recognition method based on deep belief network. *Chin. J. Petrol.* **2018**, *34*, 343–351.
12. Liu, P.; Ye, S.; Shu, Y.; Lu, X.L.; Liu, M.M. Research on coal mine safety knowledge graph construction and intelligent query method. *Chin. J. Inf.* **2020**, *34*, 49–59.
13. Hochreiter, S.; Schmidhuber, J. Long short-term memory. *Neural Comput.* **1997**, *9*, 1735–1780. [CrossRef] [PubMed]
14. Devlin, J.; Chang, M.-W.; Lee, K.; Toutanova, K.J.a.p.a. Bert: Pre-training of deep bidirectional transformers for language understanding. *arXiv* **2018**, arXiv:1810.04805.
15. Mikolov, T.; Chen, K.; Corrado, G.; Dean, J. Efficient estimation of word representations in vector space. *arXiv* **2013**, arXiv:1301.3781.
16. Lample, G.; Ballesteros, M.; Subramanian, S.; Kawakami, K.; Dyer, C. Neural architectures for named entity recognition. *arXiv* **2016**, arXiv:1603.01360.
17. Viterbi, A.J. A personal history of the Viterbi algorithm. *IEEE Signal Processing Mag.* **2006**, *23*, 120–142. [CrossRef]
18. Vaswani, A.; Shazeer, N.; Parmar, N.; Uszkoreit, J.; Jones, L.; Gomez, A.N.; Kaiser, Ł.; Polosukhin, I. Attention is all you need. In *Advances in Neural Information Processing Systems 30, Proceedings of the 31st Conference on Neural Information Processing Systems (NIPS 2017), Long Beach, CA, USA, 4–9 December 2017*; Curran Associates Inc.: Red Hook, NY, USA, 2017.
19. Bahdanau, D.; Cho, K.; Bengio, Y. Neural machine translation by jointly learning to align and translate. *arXiv* **2014**, arXiv:1409.0473.
20. Sanh, V.; Debut, L.; Chaumond, J.; Wolf, T. DistilBERT, a distilled version of BERT: Smaller, faster, cheaper and lighter. *arXiv* **2019**, arXiv:1910.01108.
21. Liu, Y.; Ott, M.; Goyal, N.; Du, J.; Joshi, M.; Chen, D.; Levy, O.; Lewis, M.; Zettlemoyer, L.; Stoyanov, V. Roberta: A robustly optimized bert pretraining approach. *arXiv* **2019**, arXiv:1907.11692.
22. Yang, J.; Zhang, Y.; Li, L.; Li, X. YEDDA: A lightweight collaborative text span annotation tool. *arXiv* **2017**, arXiv:1711.03759.

23. Chen, Z.L.; Yuan, F.; Li, X.H.; Zhang, M.M. Joint extraction of named entities and relations from Chinese rock description text based on BERT-BiLSTM-CRF Model. *Geol. Rev.* **2022**, *68*, 742–750. Available online: https://kns.cnki.net/kcms/detail/detail.aspx?doi=10.16509/j.georeview.2022.01.115 (accessed on 23 July 2022).
24. Xie, T.; Yang, J.A.; Liu, H. Chinese entity recognition based on BERT-BiLSTM-CRF model. *Comput. Syst. Appl.* **2020**, *29*, 48–55.

applied sciences

MDPI

Article

Mineral Identification Based on Deep Learning Using Image Luminance Equalization

Junyu Zhang [1,2], Qi Gao [1,2], Hailin Luo [1] and Teng Long [1,2,*]

[1] School of Information Engineering, China University of Geosciences, Beijing 100083, China; 1004196122@cugb.edu.cn (J.Z.); gaoqi1024@cugb.edu.cn (Q.G.); 15779733478@139.com (H.L.)
[2] Key Laboratory of Network Assessment Technology, Institute of Information Engineering, Chinese Academy of Sciences, Beijing 100093, China
[*] Correspondence: longteng@cugb.edu.cn

Abstract: Mineral identification is an important part of geological research. Traditional mineral identification methods heavily rely on the identification ability of the identifier and external instruments, and therefore require expensive labor expenditures and equipment capabilities. Deep learning-based mineral identification brings a new solution to the problem, which not only saves labor costs, but also reduces identification errors. However, the accuracy of existing recognition efforts is often affected by various factors such as Mohs hardness, color, picture scale, and especially light intensity. To reduce the impact of light intensity on recognition accuracy, we propose an efficient deep learning-based mineral recognition method using the luminance equalization algorithm. In this paper, we first propose a new algorithm combining histogram equalization (HE) and the Laplace algorithm, and use this algorithm to process the luminance of the identified samples, and finally use the YOLOv5 model to identify the samples. The experimental results show that our method achieves 95.6% accuracy for the identification of 50 common minerals, achieving a luminance equalization-based deep learning mineral identification method.

Keywords: deep learning; image enhancement; mineral identification; convolutional neural networks

Citation: Zhang, J.; Gao, Q.; Luo, H.; Long, T. Mineral Identification Based on Deep Learning Using Image Luminance Equalization. *Appl. Sci.* **2022**, *12*, 7055. https://doi.org/10.3390/app12147055

Academic Editor: Andrés Márquez

Received: 13 June 2022
Accepted: 10 July 2022
Published: 13 July 2022

1. Introduction

Mineral identification occupies an important position in geological research. Traditional geological mineral identification methods mainly identify minerals by the naked eye or observation instruments. Naked-eye identification heavily relies on the discriminatory ability of the identifier. Observations through instruments, such as the identification of clay minerals and hydrocarbons by using near-infrared spectroscopy [1], and mineral identification and mineral mapping by imaging spectroscopy [2], require special identification instruments. Both methods are labor intensive and their accuracy is often influenced by the experience and ability level of the identifier. In recent years, researchers have used deep learning techniques to reduce these effects, for example, Porwal et al. [3] used artificial neural networks in mineral potential mapping, and Li et al. [4] used convolutional neural networks based on geological big data for mineral prospect prediction. In mineral identification, many works also use intelligent algorithms, and these methods can be classified into three categories according to the test method and the type of data obtained: identification based on chemical composition analysis; identification based on spectral analysis; and identification based on optical pictures. The main types of data involved in identification methods based on chemical composition analysis [5] are energy scattering spectroscopy (EDS) [6], electron probe (EPMA) [7], and laser-induced breakdown spectroscopy (LIBS) [8]. The identification method [9] based on spectral analysis is the most reliable method for mineral identification, but it requires expensive testing instruments and is therefore difficult to be widely promoted. The optical picture-based identification method is the most common identification method, which can be performed by microscopic images [10–15] and ordinary photographs [16–18]. As shown in Table 1, we summarized the different current mineral identification methods.

All of the above studies enable the identification of minerals, but usually only for a small number of species of minerals, and also lack stable and excellent identification accuracy. In addition, one difficulty in using photo-based mineral identification is that mineral photos in the field are often affected by light intensity as well as shadows, resulting in photos with different photometric details, which can easily lead to errors in identification. For example, the same mineral may be pictured in two colors with strong and weak light intensity, and color is one of the important features used for mineral identification. Therefore, it is difficult to achieve high accuracy with a direct identification of photos taken with cell phones or cameras. Studies using image enhancement techniques to eliminate the effects of extraneous factors on photographs have emerged and demonstrated utility in many other applications. For example, Zhi et al. [19] investigated a new method to improve the change detection accuracy of synthetic aperture radar (SAR) remote sensing images by combining image enhancement algorithms based on wavelet and spatial domains and power law. In addition, regarding the effect of luminance, Xiao et al. [20] relied on Retinex theory and used a two-step approach combining candidate regions and object locations to achieve object recognition in low luminance situations. Xiong et al. [21] achieved the identification of ripe litchi under different lighting conditions based on Retinex image enhancement and improved the accuracy of image identification. In more detail, we compare the accuracy of mineral identification approaches based on image type later on, as shown in Section 4.3.

Table 1. Comparison of different mineral identification methods.

Methods	Studies	Characteristics
Instrument Observation	[1]	Wide range of applications.
	[2]	Spectrometer with very high pixels.
Chemical Composition Analysis	[6]	Fast data acquisition.
	[7]	High accuracy of chemical element identification.
	[8]	Low sample loss.
Spectral Analysis	[9]	Reliable and has international datasets.
Micro-optical Picture Analysis	[10]	High accuracy rate.
	[11]	Effectively differentiate between quartz and resin.
	[12]	Effective mineral grain identification.
	[13]	Good results for rock minerals.
	[14]	High accuracy of sulfide mineral identification.
	[15]	Good performance in petrographic thin sections.
Traditional Image Analysis	[16]	Combined with mineral hardness.
	[17]	High accuracy of malachite and blue copper mineral identification.
	[18]	Be able to distinguish the formation minerals of different granite types.

There are many models for object detection, such as EfficientDet [22] and YOLOv5. YOLOv5 (as shown in Section 3) extends from YOLOv4 [23], which is one of the most effective object detection models available. Yolov5 has been used in many practical applications such as face recognition [24] and aircraft target detection [25].

In this paper, we combine image enhancement techniques with YOLOv5 for mineral detection to address the effects of illumination factors on image chromatic aberrations. With this method, we achieved the accurate identification of mineral images without relying on specialized instruments for obtaining identification data. In addition, our method enables the more accurate identification of samples with poor lighting conditions (too bright or too dark) than other efforts to identify minerals based on image data. Moreover, our work expands the range of mineral species that can be identified to a greater extent than other works. Our detailed contributions are shown below.

- We first propose a novel image enhancement algorithm, one which combines histogram equalization (HE) and the Laplace algorithm. In subsequent experiments, the algorithm shows powerful results.

- We achieved the an efficient identification of 50 minerals, which is a significant expansion of the number of mineral species identified compared to the existing works.
- Experiments show that our method achieves 95.6% accuracy in mineral identification, surpassing existing mineral identification methods.

The content of this paper is shown as follows. We introduce a novel image enhancement approach in Section 2, combining histogram equalization (HE) and the Laplace algorithm. In Section 3, we focus on the structure of the model we use and briefly describe the training environment and process. In Section 4, we show the results of our experiments and compare them with other methods, in addition to evaluating the model effectiveness using objective evaluation metrics. In Section 5, we conclude the article and propose future work.

2. The Proposed Method

2.1. Histogram Equalization

Histogram equalization [26] is an important method for the statistical analysis of the image grayscale distribution and is useful for images where both the background and foreground are too bright or too dark. This method enables more detail in overexposed or underexposed [27] photographs. The traditional histogram equalization method uses the cumulative distribution function of the probability of each gray level of the image as the transformation function, and according to this transformation relationship, an image with uniformly distributed gray probability density can be obtained. Its cumulative distribution function can be expressed as:

$$s_k = T(r_k) = \sum_{j=0}^{k} \frac{n_j}{n} = \sum_{j=0}^{k} p_r(r_j) \quad , \quad 0 \leqslant r_j \leqslant 1, \quad k = 0, 1, \cdots, L-1 \tag{1}$$

where r_j is the normalized gray level before the transformation, $T(r_k)$ is the transformation function, s_k is the normalized gray level after the transformation, n_j is the number of pixels with the k-th gray level in the original image, n is the total number of pixels in the image, and $p_r(r_j)$ is the probability of taking the k-th gray level in the image before the transformation. However, due to its unselective data processing, it may increase the contrast of background noise and decrease the contrast of useful signals. In addition, the gray level of the transformed image is reduced and some details may be lost. Some images, such as histograms with peaks, are processed to show the unnatural over-enhancement of contrast.

2.2. Laplace Operator Image Enhancement

The Laplace operator [28] image enhancement is widely used in image processing as a second-order differential algorithm commonly used in the field of digital image processing. It causes the gray contrast to be enhanced, thus making the blurred image sharper. The essence of image blurring is that the image is subject to averaging or integration operations, so the image can be inverse operated. For example, differential operations can highlight image details and make the image sharper. Since Laplace is a differential operator, its application enhances the areas of sudden gray changes in the image and attenuates the areas of slow gray changes. Therefore, the Laplace operator can be selected to sharpen the original image to produce an image describing the abrupt grayscale changes, and then the sharpened image is produced by superimposing the Laplace image with the original image. The basic method of Laplace sharpening can be represented by the following equation.

$$L(x,y) = \begin{cases} f(x,y) - \nabla^2 f(x,y), t \leqslant 0 \\ f(x,y) + \nabla^2 f(x,y), t > 0 \end{cases} \tag{2}$$

where $f(x,y)$ denotes the two-dimensional image, $\nabla^2 f(x,y)$ denotes its Laplace operator, and t is the neighborhood center comparison coefficient. This simple sharpening method produces the effect of a Laplace sharpening process while preserving the background information. By superimposing the original image to the processing result of the Laplace transform, we can preserve each gray value in the image so that the contrast at the gray abrupt change is enhanced. The final outcome is to bring out small details in the image while preserving the image background. However, this tends to produce a double response to image edges, which will affect the experimental results.

2.3. A New Algorithm Based on HELaplace

In order to overcome the shortcomings of the aforementioned classical histogram and Laplace algorithms, and considering the characteristics of using image fusion, this paper proposes a new algorithm for image enhancement by HELaplace. In this paper, we combine the idea of image fusion by first processing the images with histogram equalization algorithm and Laplace operator, respectively, and then fusing the processed images into a new image after weighting the average by a certain proportion. This approach demonstrates a good enhancement effect within a certain percentage range.

We convert the input image G into YCrCb (a kind of color coding method) [29] space, and then separate the YCrCb image channels and equalize the image histogram using the CLAHE [30] algorithm, which can improve the details of the image while avoiding the problem of the excessive contrast enhancement of the image. The processed channel and the unprocessed channel are combined and then converted to RGB image A. The image is then sharpened and enhanced using the 8-neighborhood Laplace operator with center 5 and image convolution, and the enhanced image is noted as B. The weighted average image fusion algorithm can be expressed as:

$$F(i,j) = \lambda A(i,j) + (1-\lambda)B(i,j) \tag{3}$$

where the input image $A(i,j)$ represents the illumination function of the image after HE algorithm processing, $B(i,j)$ represents the illumination function of the image after Laplace processing, and the output image $F(i,j)$ represents the fused image. The size of the image is 256 × 256 pixels, i and j are the coordinates of a pixel in the image, and $i, j \in [256, 256]$, $A, B \in [0, 255]$.

The algorithm description of HELaplace is shown in Algorithm 1. We apply the HELaplace algorithm to the same image and the result is shown in Figure 1. By comparison, we can see that the image is better after the HELaplace algorithm.

Algorithm 1 HELaplace

Input: Original image G
Output: Synthetic image G_0

YCrCb = COLOR_BGR2YCR_CB(G)
// Converting RGB images to YCrCb space
channels = split(YCrCb)
// Separate YCrCb image channels
YCrCb = merge(channels)
// Convert YCrCb image back to RGB image
G_1 = COLOR_YCR_CB2BGR(YCrCb)
kernel = np.array
// np.array is the Laplace operator
G_2 = filter2D(G_1, cv2.CV_8UC3, kernel)
// Convolve G with kernel
G_0 = a*G_1+b*G_2
// a,b are coefficients

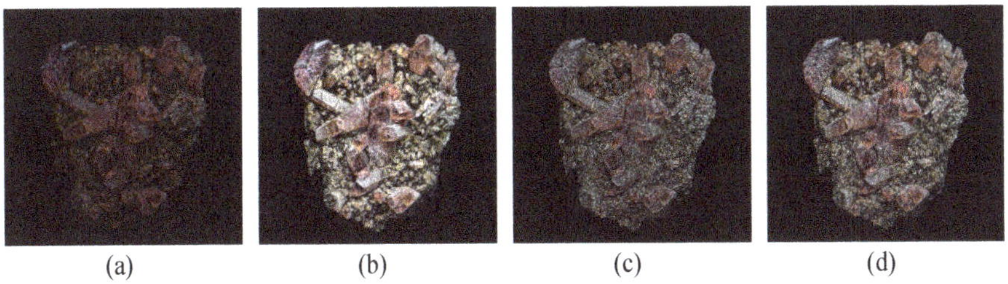

Figure 1. (**a**) Original image; (**b**) image processed by HE; (**c**) image processed by Laplace; and (**d**) image processed by HELaplace.

3. Architecture of the Neural Network

3.1. Description of Our Model

The main procedure of the experiment is shown in Figure 2. First, we collect data on a variety of minerals. Then, we label all the data and split the dataset into a training set and a test set. HELaplace processing is performed on the data from the test set and training set. Then, the obtained training set is used to train in a convolutional neural network through the YOLOv5 model. Finally, the classification to which each mineral picture in the test set belongs is calculated and the accuracy rate is recorded.

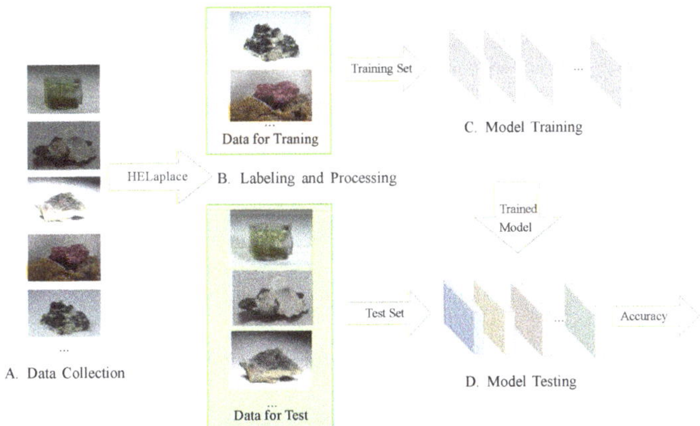

Figure 2. The structure of our model.

Specifically, Figure 3 illustrates the specific structure of the YOLOv5 network. It consists of four parts: input, backbone, neck, and prediction. The input side uses Mosaic data enhancement [23] and adaptive anchor frame calculation. The backbone part uses the focus structure and the cross-stage-partial-connections (CSP) structure. The neck part uses a feature pyramid network (FPN) + pixel aggregation network (PAN)) structure. The prediction part uses non-maximal suppression (NMS) to filter the targets, so it has high accuracy. As a new type of deep neural network (DNN), unlike traditional algorithms that require strict image pixel size, YOLOv5's adaptive image scaling has no requirement in terms of image size. We also modified the YOLOv5 code in the letterbox function of datasets.py to add a minimum of black borders to the adaption of the original image, reducing information redundancy and therefore greatly improving the processing speed. The CSP structure of YOLOv5s is to divide the original input into two branches and perform separate convolution operations to halve the number of channels. One branch performs the Bottleneck * N operation, then concats two branches. This allows the input and output of BottlenneckCSP to be the same size, which enables the model to learn more features. The neck of YOLOv5 has the same FPN+PAN structure as in YOLOv4. However, the convolution operation used in the neck of YOLOv4 is regular. In contrast, the CSP2 structure inspired by the CSPNet [31] design is used in the neck structure of YOLOv5 to enhance the network feature fusion and improve the identification accuracy.

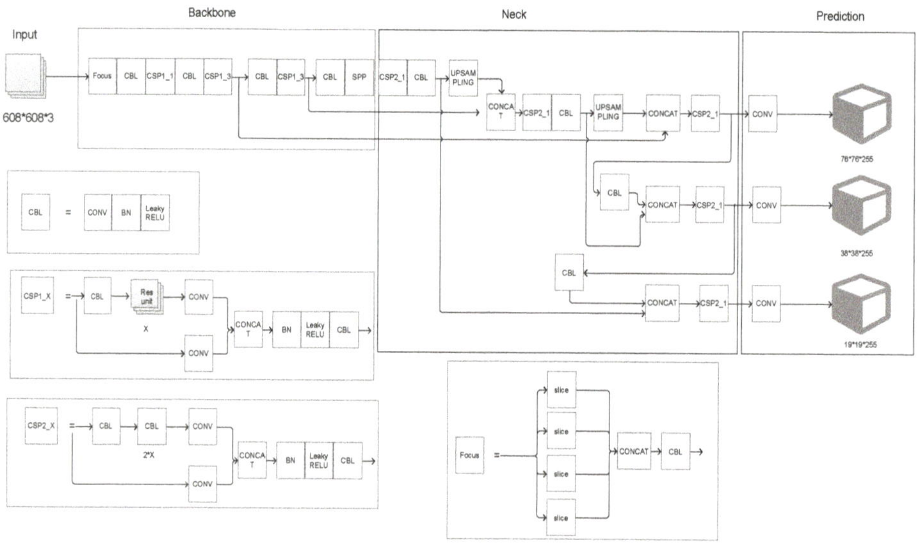

Figure 3. The main model of YOLOv5 network.

3.2. *Model Training*

In this paper, the deep learning integrated development environment is Pycharm. Test environment: NVIDIA GTX 1060, 8G memory, Intel Core(TM) i7-8750H CPU, and Python 3.8 as the compiler language. The parameters we used for model training are shown in Table 2. Parameters not listed in the table are used as default values.

Table 2. Parameters used for model training.

Parameters	Configuration
Pre-training weight	YOLOV5S.PT
Epochs	100
Sample size	183,380
Conf-thres	0.05
Iou-thres	0.45
Img-size	640
Batch-size	10

In our experiments, we use the GLOU function [32] as our loss function. Its smaller value indicates more accurate results. The expression of its function is

$$\text{Glou}_{\text{loss}} = -\frac{1}{\sum_p 1}\sum_p (1 - \text{lou}_p) \tag{4}$$

where p denotes the predicted positive example index and lou_p is the intersection ratio of the predicted positive example frame p to the corresponding true frame.

We recorded the changes in loss function GLOU values during the training process and tested the accuracy of the model on the validation set after each iteration of the training set was completed. The change in GLOU loss during the training process is shown in Figure 4. It can be seen that the model converges effectively, and the GLOU loss has reached a low level after 50 epochs. According to the figure, the model achieves the best accuracy on the validation set after the 90th iteration, and the accuracy decreases after continuing the training, probably due to some overfitting.

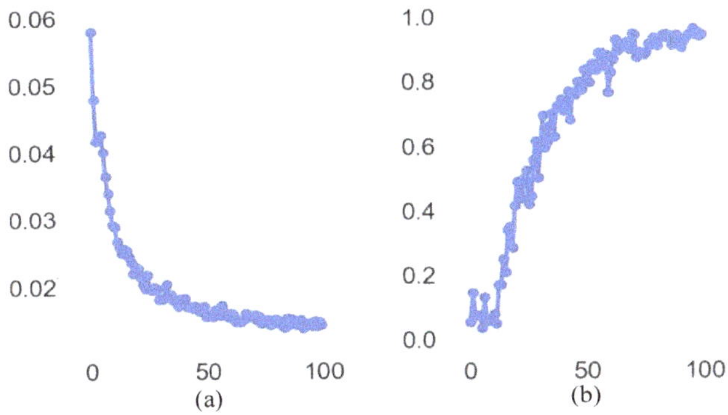

Figure 4. Change in (**a**) loss value and (**b**) precision value.

4. Test Result and Discussion

To test the accuracy of our method, we selected 13,911 images from a collection of 220,057 images for testing our neural network model. After inputting one of the images into the neural network, the mineral category with the highest probability is given. We evaluate the performance of our method in terms of accuracy, and also compare it with other methods and give results.

4.1. Data

The training of mineral identification using YOLOv5 requires a large amount of data during validation and testing. The more data available for training, the more generalizable and robust the model will be, and the higher the accuracy will be. To obtain a large amount of specialized image data for a wide range of minerals, we chose to use image data from Mindat [33]. Mindat is a community-led global mineral and provenance database website and the world's largest database of mineral information. In this paper, one mineral is selected as a training representative in the database according to the mineral category criteria [33] in each mineral major category to obtain adequate category coverage. To further extend the mineral coverage categories, we expanded 26 minerals from those covered by work [16]. Therefore, the images of a total of 50 minerals were collected as experimental samples. The names of relevant minerals and the number of samples are shown in Table 3. Among them, the small numbers of samples of certain minerals are due to their rarity, which makes it difficult to obtain a large number of samples. It is worth noting that all samples of minerals in this paper are labeled according to the classification criteria of Mindat.

Since some of the images directly obtained from the website were taken under a microscope or after processing, this may have some influence on the experimental results. Therefore, we artificially removed the images that did not meet the requirements in the dataset during the collection process. We uniformly mixed each of the obtained mineral images in the ratio of 10:1:1 and separated them into a training set, a validation set, and a test set. An example of the mineral images is shown in Figure 5.

Table 3. Names of the minerals and the number of samples.

#No.	Mineral	Number of Samples	#No.	Mineral	Number of Samples
1	Adularia	738	26	Hematite	6086
2	Aegirine	909	27	Magnetite	2615
3	Agate	3636	28	Malachite	7919
4	Albite	1882	29	Marcasite	1748
5	Almandine	2124	30	Moissanite	10
6	Amber	294	31	Niccolite	245
7	Anglesite	1981	32	Nitratine	10
8	Azurite	8320	33	Opal	3283
9	Beryl	9836	34	Orpiment	754
10	Biotite	1437	35	Ozocerite	23
11	Boracite	240	36	Pyrite	13,042
12	Cassiterite	3321	37	Quartz	46,398
13	Chalcopyrite	3296	38	Rhodochrosite	4510
14	Cinnabar	1618	39	Ruby	872
15	Copper	5504	40	Sapphire	1056
16	Demantoid	785	41	Schorl	2200
17	Diopside	1649	42	Selenium	106
18	Elbaite	5683	43	Sphalerite	6412
19	Epidote	3915	44	Stibnite	2548
20	Fluorite	28,147	45	Sulphur	1843
21	Galena	6661	46	Topaz	3926
22	Goethite	4063	47	Torbernite	1170
23	Gold	4796	48	Turquoise	988
24	Gypsum	2439	49	Whewellite	94
25	Halite	821	50	Wulfenite	8104
	Total			220,057	

Figure 5. Examples of cropped images ((**a–d**) chalcopyrite; (**e–h**) copper; (**i–l**) elbaite; and (**m–p**) demantoid). The original images are from Mindat.

4.2. Test Result

We used the YOLOv5 neural network and HELaplace+YOLOv5 neural network to test images with too little light and too much light, respectively, and the average accuracy obtained is shown in Table 4. The test results show that the combination of the HELaplace and YOLOv5 algorithms can greatly improve the identification accuracy. Figure 6 shows the accuracy of mineral identification for all 50 categories. As we can see, except for specific minerals, all of our minerals are identified with an accuracy of more than 80%. Among them, four minerals possess relatively low accuracy due to the a small number of training samples, which include Moissanite, Nitratine, Ozocerite and Selenium. Using HELaplace in combination with YOLOv5, the accuracy of all mineral species was improved compared to the results without using HELapace, especially the identification accuracy of minerals (Azurite, Chalcopyrite, Galena, Topaz) which was improved by 10%. The main reason is that the images taken in insufficiently or excessively bright light will have chromatic aberrations due to the light, many minerals have similar shapes and textures, and the resulting chromatic aberrations make it difficult for the model to correctly identify them based on the images. After applying HELaplace, the minerals (Adularia, Magnetite, and Malachite) do not significantly improve the accuracy, which is due to the fact that these minerals themselves are too dark and less influenced by light. It can be seen from Table 4 and Figure 6 that combining HELaplace with YOLOv5 can improve the identification accuracy of most minerals.

Table 4. Comparison of the accuracy of different methods.

Method	Accuracy
YOLOv5	85.31%
HE + YOLOv5	87.14%
Laplace + YOLOv5	86.82%
HELaplace + YOLOv5	95.63%

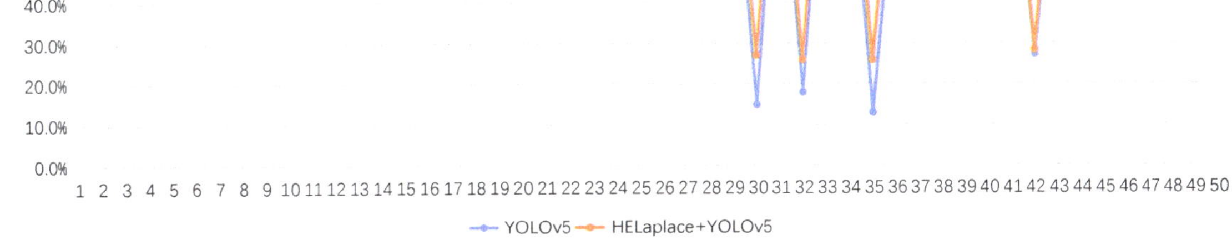

Figure 6. Accuracy comparison of specific mineral species.

4.3. Comparison with Other Methods

Table 5 demonstrates the number and accuracy of identified minerals for existing mineral detection methods. In contrast to the dual-energy CT chemometric calibration method [34], our work does not require the use of instruments for the medical X-ray tomography of minerals. Compared to the method using polarized light microscopy to obtain images [10], which can only identify five minerals, our method can identify 50 species with similar accuracy. Similarly, compared to the work of Julio et al. [11] which could only distinguish between resin and quartz, we were able to differentiate more minerals and maintain a similar accuracy. Furthermore, in contrast, we do not need special instruments to obtain the picture data under the polarized light microscope, we only need to take pictures of the minerals to perform the identification. In contrast to the work of Zeng et al. [16] who used Mohs hardness and images to identify minerals, our work does not require the use of instruments to obtain Mohs hardness. It is worth noting that the aforementioned work on identification using mineral images was experimentally measured using images taken under normal lighting, and when experiments were conducted using images taken under excessively dark or excessively bright conditions, the accuracy rate would be reduced to varying degrees.

Table 5. Comparison of deep learning-based and image-based mineral identification works.

Studies	Accuracy (%)	Number of Identified Minerals	Image Type
[10]	89	5	Microscopic
[11]	95	2	Microscopic
[12]	90	9	Microscopic
[13]	90.9	4	Microscopic
[14]	90	4	Microscopic
[15]	95.4	23	Microscopic
[34]	\	23	CT
[35]	94.2	5	Raman spectra
[16]	90.6	36	Photo and hardness
[17]	86	16	Photo
[18]	90	7	Photo
Our method	95.6	50	Photo

4.4. Objective Evaluation Indicators

Since it is difficult to obtain the normal illumination image corresponding to the image under abnormal illumination, for the image quality after enhancement, natural image quality evaluator (NIQE) [36] was used in this paper. NIQE is a non-reference image quality index often used to measure the quality of the image, a smaller NIQE indicating a better the quality of the measured picture. In addition, we used the lightness-order-error (LOE) [37] to evaluate the contrast of the enhanced image with the original illuminated image. LOE reflects the natural retention of the image, and a smaller value indicates that the image has a better order of luminance and therefore looks more natural. Table 6 shows the objective evaluation data of the corresponding methods in Figure 1. From the data in the table, we can see that the LOE of our algorithm is lower than that of the Laplace algorithm, and it is the lowest among all algorithms, indicating that we have the best result in maintaining the naturalness of the image. Furthermore, the NIQE value of the algorithm in this paper is the lowest among all algorithms, which indicates that the method in this paper does not produce much detail, thus blurring and color distortion to the original image.

Table 6. Result of LOE and NIQE.

Index	HE	Laplace	HELaplace
LOE	222.6444	156.2836	150.7435
NIQE	25.3780	41.7903	25.2050

5. Conclusions and Future Work

In this paper, we propose a deep learning mineral identification method based on luminance equalization. Compared with traditional mineral identification methods, we reduce the reliance on the researcher's experience and instruments. Compared with traditional mineral identification algorithms, we reduce the influence of illumination intensity on mineral identification and greatly improve the accuracy rate. In the deep learning recognition part, we used YOLOv5 to further improve the identification accuracy. During model selection, we used the optimized YOLOv5 to further improve the identification accuracy. In the future, more features will be introduced, such as combining the density and transparency of minerals with photos to further improve the accuracy of mineral identification. However, the identification method mentioned in this paper has some limitations: when the input picture is a mineral other than fifty minerals, the closest one among fifty minerals will be given. In the future, we will collect more mineral data to address this issue.

Author Contributions: Conceptualization, T.L.; Data curation, J.Z. and H.L.; Funding acquisition, T.L.; Methodology, J.Z.; Supervision, T.L.; Visualization, J.Z., Q.G. and H.L.; Writing—original draft, J.Z. and Q.G.; Writing—review and editing, T.L. All authors have read and agreed to the published version of the manuscript.

Funding: This work was supported by National Natural Science Foundation of China under Grants No. 62002332, 62072443.

Institutional Review Board Statement: Not applicable.

Informed Consent Statement: Not applicable.

Data Availability Statement: Data sharing is not applicable.

Acknowledgments: Thanks are given to Zhi Wang, Zhujun Nie, and Zexin Wu for their help in collecting the mineral samples.

Conflicts of Interest: The authors declare no conflict of interest. The funders had no role in the design of the study; in the collection, analyses, or interpretation of data; in the writing of the manuscript, or in the decision to publish the results.

References

1. Lu, Y.; Yang, K.; Xiu, L. Identification of hydrocarbon and clay minerals based on near-infrared spectroscopy and its geological significance. *Geol. Bull. China* **2017**, *36*, 1884–1891.
2. Wang, R.-S.; Yang, S.-M.; Yan, B.K. A review of mineral spectral identification methods and models with imaging spectrometer. *Remote Sens. Land Resour.* **2007**, *19*, 1–9.
3. Porwal, A.; Carranza, E.; Hale, M. Artificial neural networks for mineral-potential mapping: A case study from Aravalli Province, Western India. *Nat. Resour. Res.* **2003**, *12*, 155–171. [CrossRef]
4. Li, S.; Chen, J.; Liu, C.; Wang, Y. Mineral prospectivity prediction via convolutional neural networks based on geological big data. *J. Earth Sci.* **2021**, *32*, 327–347. [CrossRef]
5. Lou, W.; Zhang, D.; Bayless, C.R. Review of Mineral Recognition and Its Future. *Appl. Geochem.* **2020**, *122*, 104727. [CrossRef]
6. Ruisanchez, I.; Potokar, P.; Zupan, J.; Smolej, V. Classification of Energy Dispersion X-ray Spectra of Mineralogical Samples by Artificial Neural Networks. *J. Chem. Inf. Model.* **1996**, *36*, 214–220. [CrossRef]
7. Tsuji, T.; Yamaguchi, H.; Ishii, T.; Matsuoka, T. Mineral classification from quantitative X-ray maps using neural network: Application to volcanic rocks. *Island Arc* **2010**, *19*, 105–119. [CrossRef]
8. El Haddad, J.; de Lima Filho, E.S.; Vanier, F.; Harhira, A.; Padioleau, C.; Sabsabi, M.; Wilkie, G.; Blouin, A. Multiphase mineral identification and quantification by laser-induced breakdown spectroscopy. *Miner. Eng.* **2019**, *134*, 281–290. [CrossRef]
9. Liu, J.; Osadchy, M.; Ashton, L.; Foster, M.; Solomon, C.J.; Gibson, S.J. Deep convolutional neural networks for raman spectrum recognition: A unified solution. *Analyst* **2017**, *142*, 4067–4074. [CrossRef]
10. Guo, Y.; Zhou, Z.; Lin, H.; Liu, X.; Chen, D.; Zhu, J.; Wu, J. The mineral intelligence identification method based on deep learning algorithms. *Earth Sci. Front.* **2020**, *27*, 39–47.
11. Álvarez Iglesias, J.C.; Santos, R.B.M.; Paciornik, S. Deep learning discrimination of quartz and resin in optical microscopy images of minerals. *Miner. Eng.* **2019**, *138*, 79–85. [CrossRef]
12. Maitre, J.; Bouchard, K.; Bédard, L.P. Mineral grains recognition using computer vision and machine learning. *Comput. Geosci.* **2019**, *130*, 84–93. [CrossRef]
13. Zhang, Y.; Li, M.; Han, S.; Ren, Q.; Shi, J. Intelligent Identification for Rock-Mineral Microscopic Images Using Ensemble Machine Learning Algorithms. *Sensors* **2019**, *19*, 3914. [CrossRef]
14. Xu, S.; Zhou, Y. Artificial intelligence identification of ore minerals under microscope based on deep learning algorithm. *Acta Petrol. Sin.* **2018**, *34*, 3244–3252.
15. Izadi, H.; Sadri, J.; Mehran, N.A. Intelligent mineral identification using clustering and artificial neural networks techniques. In Proceedings of the Conference on Pattern Recognition and Image Analysis, Birjand, Iran, 6–8 March 2013; pp. 1–5.
16. Zeng, X.; Xiao, Y.; Ji, X.; Wang, G. Mineral Identification Based On Deep Learning That Combines Image And Mohs Hardness. *Minerals* **2021**, *11*, 506. [CrossRef]

17. Peng, W.; Bai, L.; Shang, S.; Tang, X.; Zhang, Z. Common mineral intelligent recognition based on improved InceptionV3. *Geol. Bull. China* **2019**, *38*, 2059–2066.
18. Ramil, A.; López, A.; Pozo-Antonio, J.; Rivas, T. A computer vision system for identification of granite-forming minerals based on RGB data and artificial neural networks. *Measurement* **2018**, *117*, 90–95. [CrossRef]
19. Li, Z.; Jia, Z.; Yang, J.; Kasabov, N. A method to improve the accuracy of SAR image change detection by using an image enhancement method. *ISPRS J. Photogramm. Remote Sens.* **2020**, *163*, 137–151. [CrossRef]
20. Xiao, Y.; Jiang, A.; Ye, J.; Wang, M.W. Making of Night Vision: Object Detection Under Low-Illumination. *IEEE Access* **2020**, *8*, 123075–123086. [CrossRef]
21. Xiong, J.; Zou, X.; Wang, H.; Peng, H.; Zhu, M.; Lin, G. Recognition of ripe litchi in different illumination conditions based on Retinex image enhancement. *Trans. Chin. Soc. Agric. Eng.* **2013**, *29*, 170–178.
22. Tan, M.; Pang, R.; Le, Q.V. Efficientdet: Scalable and efficient object detection. In Proceedings of the IEEE/CVF Conference on Computer Vision and Pattern Recognition, Seattle, WA, USA, 13–19 June 2020; pp. 10781–10790.
23. Bochkovskiy, A.; Wang, C.Y.; Liao, H.Y.M. Yolov4: Optimal speed and accuracy of object detection. *arXiv* **2020**, arXiv:2004.10934.
24. Xu, Q.; Zhu, Z.; Ge, H.; Zhang, Z.; Zang, X. Effective Face Detector Based on YOLOv5 and Superresolution Reconstruction. *Comput. Math. Methods Med.* **2021**, *2021*, 7748350. [CrossRef] [PubMed]
25. Li, S.; Li, Y.; Li, Y.; Li, M.; Xu, X. YOLO-FIRI: Improved YOLOv5 for Infrared Image Object Detection. *IEEE Access* **2021**, *9*, 141861–141875. [CrossRef]
26. Pizer, M.S.; Amburn, P.E.; Austin, D.J.; Cromartie, R.; Geselowitz, A.; Greer, T.; Romeny, T.H.B.; Zimmerman, B.J. Adaptive histogram equalization and its variations. *Comput. Vis. Graph. Image Process.* **1987**, *39*, 355–368. [CrossRef]
27. Jebadass, J.R.; Balasubramaniam, P. Low contrast enhancement technique for color images using interval-valued intuitionistic fuzzy sets with contrast limited adaptive histogram equalization. *Soft Comput.* **2022**, *26*, 4949–4960. [CrossRef]
28. Van Vliet, L.J.; Young, I.T.; Beckers, G.L. A nonlinear laplace operator as edge detector in noisy images. *Comput. Vis. Graph. Image Process.* **1989**, *45*, 167–195. [CrossRef]
29. Szedo, G. *Color-Space Converter: RGB to YCrCb*; Xilinx Corp.: San Jose, CA, USA 2006.
30. Reza, A.M. Realization of the contrast limited adaptive histogram equalization (CLAHE) for real-time image enhancement. *J. VLSI Signal Process. Syst. Signal Image Video Technol.* **2004**, *38*, 35–44. [CrossRef]
31. Wang, C.Y.; Liao, H.Y.M.; Wu, Y.H.; Chen, P.Y.; Hsieh, J.W.; Yeh, I.H. CSPNet: A New Backbone that can Enhance Learning Capability of CNN. In Proceedings of the IEEE/CVF Conference on Computer Vision and Pattern Recognition Workshops, Long Beach, CA, USA, 16–17 June 2019.
32. Rezatofighi, H.; Tsoi, N.; Gwak, J.; Sadeghian, A.; Reid, I.; Savarese, S. Generalized intersection over union: A metric and a loss for bounding box regression. In Proceedings of the IEEE/CVF Conference on Computer Vision and Pattern Recognition, Long Beach, CA, USA, 16–17 June 2019; pp. 658–666.
33. A Mineral Database. Available online: https://www.mindat.org/ (accessed on 7 June 2022).
34. Martini, M.; Francus, P.; Trotta, D.S.L.; Despres, P. Identification of Common Minerals Using Stoichiometric Calibration Method for Dual-Energy CT. *Geochem. Geophys. Geosyst.* **2021**, *22*, e2021GC009885. [CrossRef]
35. Zhang, X.; Yu, M.; Zhu, L.; He, Y.; Sun, G. Raman mineral recognition method based on all-optical diffraction deep neural network. *Infrared Laser Eng.* **2020**, *49*, 20200221-1–20200221-8.
36. Mittal, A.; Soundararajan, R.; Bovik, A.C. Making a "completely blind" image quality analyzer. *IEEE Signal Process. Lett.* **2012**, *20*, 209–212. [CrossRef]
37. Wang, S.; Zheng, J.; Hu, H.M.; Li, B. Naturalness preserved enhancement algorithm for non-uniform illumination images. *IEEE Trans. Image Process.* **2013**, *22*, 3538–3548. [CrossRef] [PubMed]

MDPI
St. Alban-Anlage 66
4052 Basel
Switzerland
Tel. +41 61 683 77 34
Fax +41 61 302 89 18
www.mdpi.com

Applied Sciences Editorial Office
E-mail: applsci@mdpi.com
www.mdpi.com/journal/applsci

www.ingramcontent.com/pod-product-compliance
Lightning Source LLC
Chambersburg PA
CBHW061557210926
51850CB00009B/2136
* 9 7 8 3 0 3 6 5 8 1 8 0 4 *